TABLE OF CONTENTS

INTRODUCTION

The modern understanding of physics tells us that energy is present in all matter. Light consists of wavelengths, and everything made out of matter consists of small particles, including the human body and its energy. Some typesof particlesthat exist have a certain vibration frequency and ancient holistic medicine shares this with modern science.

Experts in crystal therapy agree that some forms of illnesses arise due to imbalances in the body. Throughout the years they have discovered that the vibration frequency of the matter inside the human body is nothing special and that there are indeed many beings that share these vibration patterns in nature.

For example, various forms of crystal are claimed to have the same energetic properties as other main energy points on the body.

Just as putting different speakers that play the same sound together can lead to an increase of the sound volume by harmonious wave amplification, a certain type of crystal which shares ourvibration frequency can increase the natural vibrations of a certain part of the body, which is suffering from some specific disease.

Crystals serve as a guide to promote the movement of natural energies into areas of the body that may be weak, and over time physical strength is re-established.

Individuals who tried this kind of holistic care reported odd feelings after having spent time in contact with their body's healing crystals. It's almost like they're instantly aware of their heartbeat, but only in the place where the stone hits. This is very odd because we normally cannot sense our heartbeat on our own.

Sources in the field of crystal healing believe this is the sensation of movement, which rises at a rate not used by the body. This is a positive indication that the natural vibration forces are returning to a normal level.

Crystal healing may also have a profound effect on the emotional condition of the patient. Mental blockages can be much harder to detect than physical ones because they do not often manifest in a way that is readily evident to someone who is not responsive to them.Nonetheless, crystal healing can provide patients with the ability to become more mentally clear, and they can easily put aside bad habits that have damaged them in the past.

While some advocates of crystal healing talk about their advantages for psychological issues, they often use it to facilitate spiritual healing. The solution to relationship problems and emotional turmoil are two of the most common applications for crystals and many can be purchased online.

The first task of someone who wants to use crystals in this way is to find the appropriate type. Many places where crystals can be ordered have dozens or even hundreds of deals.

To those who are just starting out, it can take time to sort through this huge offer. Everything from weight loss to depression is portrayed as curable by using crystals.

Most people think crystal healing will take as much as a few months to start. While some claim that it is enough to hold crystals in a pocket, others say that the effects are felt better by holding it on a chain close to the heart. Your decision will be based on your own beliefs and what you feel comfortable with.

They are also available in different sizes. Most people assume that the bigger the crystal, the greater the impact. However, with prices increasing proportionally to size, some may consider the purchasing of a smaller crystal in their first attempt.

For these purposes, the use of crystals is becoming increasingly popular among faithful people as well as non-believers, particularly when they start to catch up in celebrity circles.

This BOOK is a great resource if you are interested in learning more about how crystals can be used for healing purposes.

Let's get started!

CHAPTER 1. HISTORY AND FORMS OF CRYSTALS

What Are Crystals?

Crystals or gemstones are natural minerals or special rocks of particular characteristics. The key difference between an atom and a molecular structure is the same as that of a crystal and a typical stone.

Crystals are typically made of one or a few types of atoms that form a connected pattern of repeated atoms. A unique feature of crystals is that they grow over time. The crystal phase is called crystallization.

Crystals can absorb, store, and transmit energy, thanks to their structure. In "crystal therapy," a popular alternative treatment technique, this energy may be used for treatment. Crystal meditation incorporates the influence of crystals through the formation of crystal grids or placing them on chakras.

Crystals are wonderful 'mates,' and they greatly influence the energy of the place when used as ornamental objects. They can also boost vitality and the overall well-being of people who wear it as jewelry. Gemstones of various types emit different energy sources and should be used according to their energetic vibration and properties.

For the appropriate use of gemstones for healing and energy work, certain knowledge and practices of work with crystals, energy, and purpose must be learned. The most important things to learn are how to clean, charge, and program your crystals.

Crystal can be a strong resource for curing people with emotional, psychological, and energy illnesses, which are closely correlated with all the physical symptoms we call diseases.

History of Crystals

More than five billion years ago in a fiery supernova, a star exploded. The spinning interstellar debris created larger and larger spatial bodies over millions of years. The solar system took shape and the water vapor collected in the Earth's atmosphere as the hot red-molten rock was cooled down. It formed clouds. The primordial oceans were produced by a rain deluge; and crystals were formed.

Crystals are highly ordered, continuously moving energy units:

They grow deep inside the earth and help uncover individual hidden stocks of power and energy, metaphorically speaking. They are governed by specific mathematical rules and comply with clearly defined geometric patterns.

Under severe pressure, they are forced to transmute and are exquisitely produced by extremes of time and temperature. Each crystal is special and its features are as diverse as its strengths. The study of crystals would take years, but all crystals have unexpected advantages.

Growing crystal can be programmed for receiving, processing, releasing, reflection, refraction, magnification, transformation, balance, and harmonization, organizing, amplifying, concentrating, and redirecting energy. The understanding of their endless applications has therefore been historically recorded in virtually every society.

Pharaohs of ancient Egypt have been surrounded by huge amounts of gemstones, precious jewels and ornate golden statues to ensure their afterlife enjoyment. Such metals inlaid with gold were decorated with:

- Carnelian-believed to facilitate the passage of the soul to the netherworld
- Lapis lazuli – ancient heavenly Alchemist's Stone; used in cosmetic and painting by royal morticians
- Turquoise-thought to fend off the "evil eye"
- Quartz-the most powerful and multipurpose stone in the crystal-world
- Amber-believed to aid the exploration of ancient knowledge and wisdom

- Emeralds – Pharaohs for their incredible green colors; life, development, fertility, and innovation symbolized
- Ruby-considered one of the precious gems of our world; symbolizes power, passion, and desire
- Sapphire-the success stone and the Ruby's sister
- Topaz, once thought to be tainted by the Egyptian sun god Ra's golden glow; believed to improve physical strength

In 2950 B.C., the Chinese jade was long believed to be a sign of riches and honor. Ancient Chinese used jade extensively in treating chronic illnesses, restoring energy balance, and enhancing the divine healing force. Jade is used in feng shui to improve one's life by obtaining positive qi (energy).

Some of the most fascinating archeological findings of the 20th century are the mysteriousskulls of Mexico, Central, and South America, decorated entirely with crystals.

The mystery of the Mayan Crystal Skulls has persisted for thousands of years, especially the prophecy of the concurrent disappearance of our planet as we know it, and the great cycle of the Mayan calendar on 21 December 2012.

Individuals who are familiar with natural healing practices embrace the fundamental laws of the concepts of mind/body/spirit interaction which are central to Indian Ayurvedic and Native American systems.

Crystal healing discusses the basic concepts and characteristics of the field of human energy:

- Barite – a sky-blue stone preferred by many Native Americans who use it in their ceremonial rituals, to turn them from physical to spiritual beings; often to inspire good dreams and recall dreams
- Emeralds – the old Aztec and Incas in South America thought that emeralds were holy; the Vedas, the ancient sacred Hindu scripts, taught that emeralds are fortunate and well-being stones
- Nebula stone — black with nebula-like markings for celestial awareness and knowledge
- Rose Mangano calcite-also called the stone Reiki; healers use it

to boost their healing strength

- Serpentine-thought to protect against the bite and sting of venomous beings;
- Shiva lingam stones – a fan of Indiana Jones has been acquainted with Indy's quest for 3 holy stones symbolizing the Hindu deity, Lord Shiva. They are a symbol of yoni or female energy to maintain a perfect balance between male and female fertility.
- Sugilite- used by healers for pain relief and rest.
- Sunstone – used for dissipating negativity and fear; radiating health, good fortune, and happiness

Crystal healing practitioners build and practice personal energy awareness and healing skills. They can redirect dangerous, destructive energy habits by "listening" and "being in tune" themselves with their crystals; rejuvenate, stabilize and align their energy.

Ancient Indian culture and many eastern cultures claim that rocks have a healing ability. The human body has seven chakras, energy centers that are situated between the base of the spine and the crown in the middle of the body.

Each chakra is associated with an endocrine gland when compared to the anatomy of the body. Energy needs to flow freely across each chakra, ensuring good health, peace of mind, and harmony.

Similarly, the endocrine gland's proper function provides for physical and mental wellbeing. Every chakra has a color and a stone. These stones are the healing stones of the chakra. The stone energy activates the chakras and equalizes the capacity for good health and well-being.

The argument that crystal healing works perfectly has always been denied by modern science. The chakras (endocrine) functions can be healed by the use of crystals. Recent research found that when the participants were in contact with stones, they felt warmth, a tingling sensation, and a feeling of health.

In-depth knowledge of the rocks, minerals, and crystals used as massage stones or healing stones has been developed by crystal healers and healing practitioners.

A spa culture has developed, where the practitioner gives massages with stones. Whether or not you are a believer, the use of stones to massage or cure or decorate always has a positive impact on an individual, as they possess various kinds of energy levels.

Is crystal healing a myth?

Dr. Christopher French, a psychologist in London, tested the effect of crystals on 80 volunteers. The volunteers were prepared for what they could expect and experience.

Some have been given fake gemstones, and some have been given true gemstones. All topics witnessed what they had been told to expect. This means that the human mind may be educated and conditioned to anticipate the outcome of the experiment/test a series of results or sensations.

How do crystals work for healing?

Our whole body expresses energy in different densities and patterns. When all energy is harmoniously modeled, we experience good health. Any disruption or imbalance in energy causes physical, mental, or emotional disorders and illnesses.

Each stone is known to have energy. The gemstones radiate energies in a particular pattern due to the subatomic particle alignment at a wavelength. Crystal healers and experts understand our body's pattern and complement it with the pattern of crystal therapy pills. A balance is generated by models that help the body unblock the chakras and allow free flow of energy.

The use of soothing gemstones is a holistic approach to good health. The use of crystal healing stones is a safe way to restore health, peace of mind, and joy without any documented side effects.

Crystals cannot heal by themselves. They can instead be used as a focus for inspiration and enlightenment; they can lead practitioners to think and live in healthier ways, and they can be used as conduits for healing energy.

Since ancient Egypt, crystals have been used to restore balance and heal ailments. The idea is that harmful energy is produced and purified by their use. By giving up toxic energy, we help our bodies heal.

In addition to information on the use of crystals in ancient Egypt, documents date from 5,000 years and also show their use in Ayurvedic and traditional Chinese medicine. Crystal therapy uses gemstones for medicinal purposes.

The stones are placed on "chakras" or body energy centers. It is supposed to improve the body by curing ailments with various types of vibration crystals. Because of the different mineral content of each crystal type, each of them produces its unique vibration.

Crystals can be used for the treatment of physical, emotional, behavioral, and spiritual conditions.You can wear crystals as an aid to healing, place them around the bathroom while you bathe, or place them near your bed while you sleep.

Chakras And Crystals

Some of the most popular yet very effective and strong healing stones for beginners are:

- Red Jasper: For reinforcing a person's essential strength and it works perfectly with the root chakra.
- Tiger's eye: Builds self-esteem and builds trust and it works perfectly with sacral chakra.
- Citrine: Increases strength of life-force gives light and pleasant vibration and it works perfectly with the solar plexus chakra.
- Rose Quartz or Aventurine:The best way to open the heart to offer happiness, peace, and attachment and it works perfectly with the Heart chakra.
- Chalcedony: Strengthens the determination and unlocks the heart and it works perfectly with the Throat chakra.
- Amethyst: Improves connections to divine energies and improves intuitive insight and it works perfectly with the third chakra of the eye.
- Clear quartz: The purest crystal with intrinsic energies to amplify and complement the cumulative work of all energy systems and chakras and it works perfectly with the chakra of the crown.

I was also intrigued by and almost amazed by crystals and gemstones. As a child, I have never collected dolls or toys. I've collected rocks! My array of rocks was very impressive; I had geodes and crystals from all over the world. I didn't collect normal things, but I don't think I was your typical kid either.

I was attracted to these gemstones for some reason. I was drawn to their sheer beauty; but besides these strange natural forms, there was something else to which I was also drawn. I believe that this was the energies that attracted these crystals to me. As an adult, I began hearing about the healing properties of crystals and eventually became a licensed crystals healer.

The use of crystals to bring beneficial changes of mind, body, and spirit is called crystal healing. In my view, crystals cured illness, treated emotional pain, helped me conquer addiction, energized me, and changed my life completely!

The growing gemstone holds its specific vibration frequencies and you can change its vibration frequency and aura by putting it on your body. Crystals act as amplifiers, too. They reinforce your goal and produce the desired result much faster.

There are many crystals available in all shapes, sizes, and colors and each has its special properties to solve different physical and emotional problems. However, you might make a mad attempt to find the specific crystals you think you need. You should do it the other way around; the crystal picks you!

If you have ever been to a crystal shop that sells crystals, just stop and go back to the very first crystal you have touched, because it was undoubtedly the crystal you wanted.

There are preparations after you buy your new crystals, which must first be completed before they are ready for use. I'm sure you're already excited and want to use them right away, but these next steps are key to the entire process of crystal healing.

You must first purify the crystals of any harmful energy they may contain. If they were in a shop, a lot of people could have touched them and the crystal could have absorbed all the negative energy they had and you do not want that.

Depending on how you use it, youneed to clean your crystals from time to time. There is a variety of efficient crystal clearing methods. The steps are discussed below:

- Step One: The simplest and most efficient approach is to put it in your hands and hold it under running water for 1-2 minutes to extract all harmful energy from the crystal. Many other cleaning methods are to place them in salt for many hours or bury for a day or more in the soil or simply by using your breath.
- Step Two: Adjust your new crystal to your vibration frequency and aim to only use it for your greater good or anybody else's. You simply have to hold the crystal in your hands, close your eyes, and set the two intentions. You can either tell them aloud

or just think about them.

- Step Three: Load your crystals. Just place your crystals for a minimum of 5 hours in direct sunlight. You can also place them under full or new moonlight overnight.
- Step Four: The last step is to configure your crystals for when you would need them. You can claim, for example, whether it is for general healing or defense, or grounding, etc... It is an optional step that can be done or not.

You have to hold the crystal in your hands and say "This crystal will be used for" then fill it in the blank.

Repeat these terms 3 to 4 times to configure the crystal fully. You're ready to use your new crystals now!

You can use your crystals in many different ways; everything depends on your needs. You can wear them as jewelry, or bring them with you in your pockets or bags (if you use them in this way, remember that you have to clean your crystals more frequently as they can more easily absorb any harmful effects in the atmosphere).

You can also meditate on or around your crystals, which significantly amplifies your meditation. By placing the corresponding stones on each chakra, you can balance and clear your chakras.

You can store crystals for clearing or protection in your room or workspace. In the night, you can also place some stones under your pillow to strengthen and recall your dreams or to sleep better. Crystals offer endless possibilities and are sure to improve your life in all ways!

Types Of Crystals

For reasons as varied as the types of crystals themselves, human beings have long been fascinated by these exquisite natural forces products. We are attracted by their elegance, their mathematical and scientific characteristics, and how they affect nature, their relation to the earth, and the universe.

Essentially, there are two types of crystals: NATURAL AND MAN-MADE.

Here we are concerned with the healing properties of natural crystals.

Crystals can transfer and change energy significantly and have been used for healing throughout human history. They may have very mild or very profound effects on the user and others around them based on the crystal form, its physical proximity, and how it is treated.

Most people who wear crystal for the first time say they feel uncomfortable. It is due to the crystal acting on the spiritual body or on the aura area, which can influence the physical health of a person if the aura or spirit energy is not stable (and has been for some time).

They can feel very tired, have headaches, or simply feel very sick or nauseous, and often a pre-existing condition gets worse. Yet all of this is temporary.The crystal starts to clear away all the negative energy in the aura of a human.

This also works on the chakras; equilibrium is restored and 'new' energy is given. Within a few weeks after wearing a crystal, people usually feel more optimistic, energetic, and physically improved. The inner cleaning of the body and mind offered by crystal may be dramatic at first.

If you intend to buy a crystal to wear, be ready to start cleaning your system for the crystal, particularly after a while. Some people begin to weep and some people feel no difference; everything depends on the person and their state of mind.

But the effects are not limited to the human body alone. A quartz crystal in the house will add fresh energy to space and give anyone in that space a positive view.

One word of caution: I never recommend putting clear quartz crystal near hyperactive children, also it can produce an unnecessary amount of highly active energy in the bedroom while you are sleeping!

Crystals are available in so many varied shapes, colors, and sizes that it can sometimes be confusing to decide to wear them, especially if you don't know the chakra system or the aura. Iwill cover every crystal in detail, most especially the most common crystals and their use.

1. Rose Quartz: This crystal is pink. A rose quartz piece is perfect

in the house on a table or a tabletop, particularly for hyperactive children or for people that are too energized. The lovely pink color heals and soothes if a person is upset.

2. Smokey Quartz: It's a spiritual and powerful stone. It's a foundation stone because excess energy can be grounded. It's a very strong aid in meditation.An example of a smokey quartz is Amethyst.Amethyst is a sacred stone too. Itbrings a strong mental balance and it is a strong healer. Amethyst is known to be the stone for addicts. This is a beautiful stone which is normally placed next to the bed, so you can sleep well and is perfect for anyone who suffers from a headache. I know people that were using sleeping pills that threw them away when they found out how amethyst could help them sleep.

3. Clear Quartz: This crystal is a great healer. This operates on all levels of mind and body and is very powerful to give you trust and minimize negativity. It will increase the natural flow of energy and function on every aspect of the auric body. However, people with excess energy can find that they can become even more hyperactive with clear quartz crystal. A crystal of another color will probably be ideal for these people.An example of a Clear quartz is the Tiger's Eye.Tiger's Eye can be brownish red, red or blue and, because all eye crystals are also grounding stones, excess energy can be absorbed by them.

4. Rainbow Quartz: These are translucent quartz holding or stuck within prisms. Rainbows select which quartz crystals to stay in. Prisms are often very small, and there may be one or more prisms of various sizes, visibility, and strength. Each small rainbow contains all the Rainbow colors. A rainbow is a spiritual link between the earthly and the Ethereal Spirit realms.

5. Tabular Quartz:These are naturally flat on both sides and are significantly wider than the other four opposite sides of the Crystal. Such unbelievably beautiful crystals have very strong and powerful vibrations, they act as bridges or lines to increase and reinforce ties between two individuals, integrate and align the Chakra energy centers, or bind two geographical points. A crystal tablet allows you to enter the angelic world through the

veil and to hear and obtain advice from your Guardian Angels.

6. Obelisk; These crystals originated in Ancient Egypt and appeared in pairs, which typically secured the temple entrances and the holy places. The obelisk symbolizes the fall of the Egyptian Gods and Goddesses from the heavens to the Earth. It's a Greek term that means needle. Normally, needles are used to pierce cloth or leather. The Obelisk's purpose is to "pierce" the veil that divides humanity from gods and deities across the realms. This is sculptured and allows external forces to enter or break up.The natural and carved crystals of the generator have the same specific characteristics and can be used to boost meditation, tools to focus your thoughts and to stimulate and intensify the energy.

7. Apex:All six sides of a crystal form an apex or point. They are useful instruments for increasing light and the energies of all crystals within a 2-foot radius.

8. Clusters: Clusters are many crystals that have formed a single unit or a small crystal community. Crystal clusters are composed by amethyst, calcite quartz, citrine, transparent quartz, and tangerine, in many sizes, forms, and a wide variety of quartz crystals. Clusters may be used to cleanse and reload smaller crystals, to protect your home or office, and to increase your meditation vibration.The shape of a crystal strengthens its natural strength, allowing those interacting with crystals to promote meditation and healing at many levels to include cellular levels.

Often when I feel spiritual energy, I keep two tiger's eye stones, one in each hand, and can feel the excess energy of my body flowing into the stones. I can still hold the stones again and draw energy from them when I need to. These are also very calming to look at and perfect for hyperactive children.

In addition to being worn, crystals may also be used for meditation. A small cluster of plain quartz or large quartz is the best form of crystal for this. I think during meditation, one should keep the crystal in the left hand but some people prefer the right hand.

Instinct is what you will obey in terms of crystals. If you believe like you have to hold some crystal, then do so. If you don't like a crystal and somebody tells you that it's for you, don't consider it. Follow your intuition always and you won't go wrong.

Crystals are beautiful. Once you start wearing one or start working on it, your entire life changes for the better. Within the crystal family, there are many different stones and colors.

I don't have room to list them all in this chapter but I will do so in the next few chapters. Nonetheless, you can visit your local crystal shop and ask for the various types of crystals. Ask the assistant for information so that you can determine which one is right for you.

Crystals are a solid substance, sometimes transparent but sometimes colored, composed of chemicals, and are both fragile and strong; mystical and earthly; functional and decorative. Crystals can be classified by their plumbing content which adds weight and softens the material compared to other materials.

Due to this versatility, certain manufacturers that deal with crystals can shape them in various forms and cut them in different ways to reflect light. This versatility includes many purposes for crystals.

Most people believe strongly in the use of crystals for healing and meditative ability in many respects. Since the ancient times, crystals have been used to facilitate spiritual, physical, and emotional well-being. As a result, many people choose to put different crystals together in their home areas as a way of fostering peace, health, and prosperity.

Moreover, other people use crystals for spiritual purposes because they offer tranquility and concentration. Others also want to wear nearby crystals just to aid in their healing abilities. Crystals not only keep their powers close to the human body but also transform an outfit and infuse a wardrobe with style and elegance.

Yet the use of crystals isn'tstoppinghere; due to their practical and brilliant decoration, crystals have been used for a long time to manufacture beautiful kitchen items such as wine glasses, serving plates and bowls, frames and vases.

Crystals of any form can be found anywhere from jewelry stores to high-end department stores. Holistic centers sell a range of crystals.Crystals are both delicate and weighty. You should consider crystals a beautiful and valuable improvement to life, whether you use them for healing, practical use, or decorative purposes.

CHAPTER 2. THE BASICS OF CRYSTAL HEALING

What is Crystal Healing?

We use crystals or gemstones for cure in crystal healing. Such crystals are either used to cure all sorts of body pain or to purify our bodies from different impurities. There are many different areas of our body known as 'chakras,' the crystal healing is achieved by placing the crystals on our body's chakras.

There are 7 chakras in our bodies where the crystals are positioned for healing. The word "chakra" means divine energy in Hindi.

These 7 chakras are known as our body's seven energy centers. The crystal used for healing comes in various colors, and these colors play a significant part in the healing process. Such chakras contribute to the flow of energy in our body and thereby keep our body healthy and free from pain and disease.

This cycle helps to control the energy flow in the body of the individual. The person who conducts a healing process will undergo crystal healing first so that his body is clean and free of all harmful energies.

Since this energy can then be transferred via the crystals from one's body to another, therefore, the person with any disease caused by the negative energy will be healed.

Crystal-healing has become much more common today, and not only do people who are sick get this procedure but also occasionally nurses and physicians have a crystal-healing session to cure their patients. Not only that the crystals should be placed on the chakras for healing, these crystals can be held at the bedside and worn as well. The crystals used vary according to the type of disease or cure the patient wants.

Nobody knows where the cure came from.

Most people have been using this practice for years. Some people had little faith in crystal healing and many of us thought it was just spiritual. But today people are gradually starting to understand the importance of crystal healing. Also, if the scientific evidence does not prove that crystal healing works, only those who have taken these sessions know its power.

Thanks to its color and form, most crystal healers use transparent quartz to work but these stones and crystals can differ depending on the chakra they place.

Is Crystal healing harmful?

Most of us see little harm in using crystals for healing. However, if a person has other diseases, and just because he practices crystal healing stops his medicine, it can be very dangerous.

It is assumed that the life of that person will get worse if the wrong energy is

passed through during the crystal healing process. Such transferred powers

are not connected to God and can, therefore, be very harmful.

Crystal Regeneration-Body and Spirit Alchemy

The awakening of mankind to the discovery of the old and forgotten healing arts such as the use of crystals is assisted by the Earth's consciousness. Crystals are members of the mineral kingdom and are thus universal energies which allow contact and synthesis for the healing and integration of these individual energies.

Each crystal contains its own "personality" and can be used especially to better explain the essence of Earth's life. Crystals are a gift of nature to humans and are present in any form, scale, color, and composition.

Each of them has a special vibration resonance because of their varying mineral contents, intrinsic geometry, and color frequency. Thus, they can be effective healing devices.

Crystal healing is a healing process, whereby crystals or gemstone are placed on or around your body, in chakras (spiritual energy centers), used as reflexology devices to stimulate points on your feet, worn, or put into and around your home to improve the energy of feng shui.

You can put the same color crystals in the same chakra to boost the flow of energy, or just sit back and meditate by holding the crystal in your hands or the third eye (6th chakra) to receive energy and messages.

It heals holistically at a physical, mental, emotional, and spiritual level, helps guide the energy flow to a particular part of the body, restores equilibrium and eventually purifies the body through the release.

The vibration energy of the human body is a dynamic electromagnetic network in which the crystals interact as they are the ideal electromagnetic conductors in nature. Crystals function by resonance, and vibrations that cause these energy centers have been shown to have a beneficial impact on our entire body system.

Many crystals have medicinal properties known to contain minerals and are used in conventional medical practices. These are piezoelectric, meaning that electricity and light often come from distortion and can produce sound waves.

For ages, shamans and crystal healers have known the ability of crystals to focus light and sound vibrations into healing rays. These are used to reorganize subtle energies and remove stress by helping them find the root of the problem.

Reverence and the use of crystals date back to civilization's dawn. Crystals have been linked for thousands of years to different parts of the body and its organs, many of which come from conventional Western and Eastern Astrology.

India's Ayurvedic records and traditional Chinese medicine claim that crystals have healing properties and this can be found in texts older than 5,000 years. The Bible refers to crystals more than 200 times.

Crystals are found in ruins of Babylon and the ancient Egyptian and Chinese tombs of rulers and many early civilizations in the world (Mayan, Aztec, Indian, African, Celtic) used quartz crystals in sacred ceremonies.

It is up to the patient to heal himself, but there are many catalytic devices, such as crystals, that can make recovery simpler and help to reconnect to the whole.

The decision to use a crystal can be as easy as following your intuition to capture the essence of the energy to which you are drawn, or refer to crystal healing indexes that link symptoms to the corresponding crystals.

These strong, crystalline entities are sacred instruments that aid us in our journeys and serve as portals for higher consciousness, personal development, harmony, sense of peace and well-being, promoting healing, and connection to the earth and beyond.

Crystal healing is an alchemical process that harmoniously mixes crystal energy with our core vibration to bring divine clarity into being.

Scientific Evidence of Crystal Healing

Crystal Healing advocates have been around for centuries. Most people assume that it's a fraud. However, a research project has identified a scientific mechanism for the transmission of radiation from crystals in the energy field of our body.

THE THIRD EYE

The ancient Third Eye of our ancestors was designed to receive external vibrations, including light, from various frequencies. It has evolved into our pineal gland over the ages. It secretes the hormone of melatonin that controls our daily lives.The research shows how our human body receives external vibrations like the calming frequencies of certain crystals.

Amethyst crystal was found to have the best measurable energy field with the most available data from the many types of crystals studied, so the majority of the evaluations were on the amethyst crystal.

Perhaps best recognized as a key to understanding science is Einstein's famous $E = MC2$ equation, which demonstrates our bodies are two sorts of matrices: one of matter (M) that is our flesh and bone, and one of energy (E) with a measurable electromagnetic field. (C is light speed.) Crystal Healing advocates call this electromagnetic field our aura.

The fundamental research goal was to find a mechanism that would allow our bodies to receive a transmission of power from a radiating crystal. Chakras and auras come into play here.

CHAKRAS

Chakra is an ancient Sanskrit term meaning 'cycle' or 'circle' which refers to 7 specific centers of energy on the body meridian. Remember that the Brow Chakra is also called the Third Eye.

Our Western philosophy typically sees Chakras as nothing more than baseless ancient myths.Many Americans oppose the notion of chakras as a "new-age movement" concept. We should note, however, that the concept of chakras goes back many millennia.

The earliest mention of Chakras comes from the Hindu scriptures of the Vedas. Evidence suggests that the Vedas were transported to ancient India as the early settlers emigrated from a lost culture. The Chakras are called "force centers" which are filled by the energy swirls in the physical body. Their energy may also radiate in the human body from these stages.

Chakras are generally recognized as symbolic, but many people believe they are real. To understand how physical chakras can be, it is important to note that energy is physical and that chakras are thought to be energy disks. Note that energy is electricity in our brainwaves and nervous systems.

The Brow Chakra, which is called the Third Eye, is believed to be the most significant from a Crystal Healing perspective. By analyzing this Third Eye, we see that our ancestors had a Third Eye that could sense vibrations from their surroundings. Some early species, such as the Mantis and Iguana, who live today, have this.

The proponents contend that Crystal Healing is a profound and gentle, very powerful healing method. The technique is to focus energy from specific crystals on the Chakras and Aura fields of the patient. The energy is concentrated at the points where the chakras and the aura may be most sensitive.

The advocates believe that our Chakras and Auras control the flow of energy in our lives.

AURAS

The aura is the electromagnetic field around the human body, any organism, and entity. This is a real and unquestionable scientific fact.

But some Crystal Healing devotees also believe that auras can show whether somebody is highly spiritual or interacts with another entity. A person's aura colors represent different aspects of the person. Keep in mind that each color is a special frequency or vibration spectrum.

Silver indicates intellect, Ruby red represents strength, Violet indicates gentleness, Rose represents devotion and Blue Sapphire indicates healing and spirituality.

Proof shows that our Third Eye (our pineal gland) can sense the strength of certain perfect or nearly perfect crystals.We have to know, to understand how this works, that our bodies are energy matrices; that amethyst crystals are energy matrices and that the two energy fields will communicate with our Third Eye.

Let us imagine the interaction of energy vibrations in our human electromagnetic field from crystals with energy vibrations. The presence of the uniform crystal vibrations helps to make the atmosphere of the body calmer.

Research work has identified a scientific mechanism to allow the energy field of our body to receive transmissions from radiating crystals.The ancient third eye of our ancestors has developed so that it can respond to external vibrations like some crystals' calming frequencies.

Does Crystal Healing Work?

Traditionally, these beautiful gems were used to relieve tension and heal the body, when they are worn as collars, bracelets, and on ankles. It isn't shocking, therefore, that some people including rich and popular ones have chosen crystal-healing, a non-invasive treatment that is focused on these sparkling and transparent gems for the restorative and therapeutic properties they have perceived.

Some people say that everything is exaggerated. But other actors-like famous actors including Shirley MacLaine and Michael York-say crystals have a magical power that is incredible and can protect and heal. They hope that these sweet little gems can do whatever they can to shield people from muggers and help them re-establish their dreams.

The explanation is simple: gemstones or crystals are places that are known as 'chakras' on body parts or have a particular life force. Our body has seven main centers of energy in matching colors, and multicolored crystals are used.

The theory behind the crystals is to get rid of all the bad negative energy in a certain location to aid healing. They get rid of the bad and heal and contribute to the good by doing so. Many crystal practitioners believe crystal therapy is like meditation, a way of concentrating on what the body wants.

This is an alternative procedure not meant to replace some type of medical treatment. One way to look at it is to focuson your hope and it will give you a stronger sense of well-being and help you cope better if you suffer from a long-term illness with frequent hospital care.

Think about the matter in this way: investing money in tiny pieces of a crystal is nothing other than a complete waste of time and a way of cashing innocent and naive (and sometimes desperate) people off, self-proclaimed crystal therapists.

Many people conclude that crystal healing works only when a doctor is involved. Although this is the most famous method, other applications can have the same effect or no effect are as follows:

- Visit a crystal healer or someone trained in crystal healing. They can place some crystals on you to relieve a certain pain.
- Wear a gemstone or crystal to keep you safe and support you as you live your life.
- Place crystal or many crystals on your bed as you sleep. You should feel recharged while you sleep.
- Put crystals in your bathwaterfor an energetic wet experience.
- Using crystals during a meditation. It can be accomplished by holding or putting a crystal next to you. Seek to meditate while concentrating on the powers of quartz.
- Use them during reflexology and/or massage
- Rub your skin with smooth crystals for maximum effect.
- Soak the stone with water and drink it
- Create a crystal essence with a soaked crystal or stone, water, and alcohol for something a little bit more enjoyable.

GUIDELINES TO OPTIMIZE YOUR CRYSTALS USAGE

Crystal therapy adherents believe that you should obey these guidelines to optimize the benefits of your crystals:

1. Clean a crystal carefully before usage. Remember that they can carry the negative energy of another human, so that is necessary. It is meaningless if you purify them but you can if you wish by using water, light, herbs, or mere sound vibration.
2. All healthy crystals must be refilled, so take a break.
3. Don't forget that you will still program and commit the new crystal when you get it, making your purpose clear. Make sure you do so before using it unless you want to remove all of its healing properties.
4. Different crystals, like humans, have different properties. Most people claim that crystals chose you for what they can give.

A friend of mine,"Lauren"narrated her experience as she accidentally stumbled upon the use of crystals with life-changing results. On her mother's insistence, she visited a crystal therapist when she was pregnant with her second child.

Upon visitation,the practitioner laid crystals on her body.Suddenly, she could sense energy rising through her hands and feetwith no physical coercion. Her feet started to tingle and she felt incredibly calm and relaxed.

Sooner, she unexpectedly found herself weeping and she let the tears go. At least she hadn't been weeping, and the emotional relief was incredible. It was amazing that she had such relief without anyone touching or talking to her.

She realized that her weeping was solelybecause she did not want to return to work. At that time, she was a teacher andwent to work full time after her older daughter was born. She felt sick and depressed eventually, and she knew that she had to go back after the baby was born and do the same thing again.

Once she got off the therapist sofa, she felt incredibly light, like a weight lifted off her shoulders. This helped hergather the courage to talk to herhusband and describe how she felt about full-time work.It would have been too hard for her to return to the same position at the same time, but she couldn't say how she felt before or just how hard she cared about it.

She now works as a crystal healer. It's crazy how someone who doesn't believe in crystals, who thought that putting them on her body would have no impact at all, would now workas a crystal healer. But she held my mind open and when she felt my energy moving it was like a surprise to her.

As with magnet therapy, crystal therapy is usually considered a non-invasive form of alternative therapy that cannot have any harmful effects.

As long as you don't give up the traditional medication or splash out all the money of your wallet on a beautiful crystal that only gathers dust on your suit.If you want to try crystal therapy, give it a shot. Yet you must have an open mind for it to work.

CHAPTER 3. CRYSTAL HEALING AND YOUR HEALTH

Over the years, crystals have been long admired for their beauty and healing properties. Many rooms at the Melbourne Art Gallery have been built to display different crystals and describe their geological origins.

A similar set is shown in the Bathurst Museum in NSW. Throughout the years, critics have been dismissive of their art of crystal healing so that it is easy to see the science of geology keeps them to devote many rooms in these museums.

There are many books on the benefits of different crystals online, so I don't want to talk about this. What I wish to explain is why crystal healing can work and try to clarify why some people feel this sense of well-being after a crystal healing session.

All around us, including you and me, is made up of those atoms or even smaller particles that oscillate, but we don't see them moving because our senses can't catch them.

These atoms are often made up of crystals and at a certain rate oscillate. In the last 50 years or more, studies at UCLA have been done to investigate this and you can find this on the Internet.

You have a special way of going, thinking, and feeling. In this world (which you know so far) no one else is a duplicate of you. You are influenced by other people's vibrations.

Do you realize that if you just take an hour to listen to a friend on the phone who feels down, you'll also feel down when they're done?

We constantly pick up the 'vibes' of others and encourage others to connect with our energy. Similarly, the crystals and everything else that surrounds us has a certain vibration that separates a rose quartz crystal from a topaz, etc.

Our vibration starts moving differently because the crystal is close to us just like other people around us would affect us. It is one of the studies I listed earlier and is readily accessible on the Internet. When you read the chapters about the qualities of each crystal,you'll find a great deal of information on how they can influence you and how each crystal can cure different emotions, etc.

Many people testify that they feel differently after they have undergone a crystal healing. There must be an explanation of why people feel special, although some people don't believe it. Combine this with the experiments and it is open to debate whether it will ever be a placebo effect.

During a crystal healing process, the healer positions different crystals on or around the body or energy field of the client as appropriate. The client normally lies on the floor or a massage table. Ideal conditions are when the room is dimly lit; a strong scent of incense or candles and soft music. The hope is for the client to relax and crystals to do their job. How much time it takes depends on the treatment.

Crystals are said to adjust the way the body energy centers (chakras) vibrate and stabilize and remove energetic and emotional barriers in the energy field and effect in the physical organization. Through unblocking the new resources, energy, health, and life can be changed and strengthened.

Crystals are a way to balance and heal the body, mind, emotions, and spirit. If it is for you or not, only your personal experience will tell. Yet don't expect just one or two sessions to perform miracles.

Your emotional baggage and depression have taken months, even years to develop and there is no magic wand to whisk it off. It's real life, not TV, unfortunately, and it takes time for recovery.

The Benefits of Using Crystals

Being grounded is a term that is heard more and more in this time of consciousness in which we live.

Clear thought and mind control involve possessing your whole being, mind, and body at the same time. When grounded, you are capable of controlling your thoughts and freeing them from any potentially unwanted and negative feelings or emotions that could distress or disintegration.

The goal of life is to flow more, to be rooted, to be present, and crystals give you the positive energy to do so. That's what grounding will do for you.

Crystals are found deep within the earth. They are very dense and deeply rooted. Some of the most coveted and deeply strong crystals come from deep in the forest. They are formed by massive quantities of pressure from above the earth.

The Earth moves its center, where there is more gravity to concentrate its power in it; it grounds closer to the middle of the world, the earth's most grounded position.

Crystals have significant calming effects on humans. Some of the benefits are listed below:

1. They display and disintegrate complex energies.
2. They block the harmful radiation of the electromagnetism, in which our use of electronic devices like phones and computers is continuously bombarded.
3. Crystals are used for self-healing in many respects. They are used in massage, Feng Shui Reiki, yoga, meditation, reflexology, and much more you can imagine.
4. They balance the chakras, whether you have either a blocked chakra or an overactive disorder that damages every chakra, there are different types of crystals that carry the equilibrium of each chakra at a certain point of the body. Growing crystal color is related to the color of each chakra or the opposite color crystal is used to match the center of the chakra.
5. Within walls, crystals can be used to intensify the body's basic

cause and reflexology.

6. Crystals have a very strong effect on the body's energy points, they dissolve the static or blocked energy flow in your body, which creates a simple natural stream.

7. For centuries they have been used for their strong strength. They put peace and rhythm back into their lives. The use of crystals brings in a higher frequency vibration that makes all aspects of your person simpler. There are thousands of crystal types used for the many positive effects on humans. When utilized in jewelry for medicinal reasons, it also has a dual function. Finding the right stones, crystals, and jewelry is a good and useful way to learn about yourself and how you can use the crystals to your benefit.

8. Crystals seem to be solid energy that can improve and balance ours. Einstein said, "Concerning matter, we have been all wrong, we know that everything goes to other things in the same neighborhood"Think of grandfather clocks, if any of them are set together in a room after a while, they start to swing together. Women in dorms tend to menstruate simultaneously. Exercising is well-documented and one of the reasons why 'crystals are helpful to us.' If we are off balance or if we want to strengthen our health, we can use the powerful strength of crystals to create greater equilibrium within our lives and our bodies.If, as Einstein said, we all transfer energy, everything can and does affect us. All matter is moving energy seen or unseen around us. Eight dimensions have been discovered and quantified by scientists and they believe there to be at least more. We can only see one of these dimensions; all other are inaccessible to the normal eye. So much is missing, but it affects us and changes our focus, feelings, emotions, and physiology.

9. Crystals help us equalize our resources. Find out what various crystals do and use them to match these energies with our world or our body.

10. Crystals soothe us in a tense situation. Play with a few crystals and see which ones make you feel soothed. Remember that's going to change from day by day, or week by

week. Even when a difficult situation occurs, Gold, amethyst, rose quartz and lepidolite are particularly useful for stress.

11. Crystals help balance our world by putting it correctly. Our emotions can contribute to this mismatch when our homes and our workplaces are out of control and severely throw us out, even to the point of illness. In the corners of your house, emeralds will secure and balance your house. The crystal will do the same in the corners of the house. Either crystal can be used, but clear quartz is possibly the best.

12. Crystals could help to bring pieces of us into ailing. Even if we are sick, the strength of our equilibrium will help to stabilize a crystal over this part and have a significant effect on its equilibrium over time.

13. Crystals help re-balance our frequencies. Every crystal has its frequency or power. They use these energies to control the frequencies inside, in, and around us.

14. Crystals help us connect to our universal energy. Gold and silver are especially good for this and the crystals are worth the money and effort, although they are hard to find.

15. Crystals help us navigate the local energy grid.

'The benefits of crystals' can be numerous, as they have a range of strong energies which can 'use' us in our lives. Crystals can be used in many ways to deal with various forms of energy and to store it. Energy only shifts levels, like songs that float around the world.

These are too numerous for the reception, reflection, refraction, amplification, concentration, transmutation, propagation, transfer, transformation, storage, stabilization, balance, and propagation.

Also, energy can be changed with crystals in hundreds of ways.

Crystal Healing Effect In Balancing Body And Mind

Crystal healing has repeatedly been reiterated that it is not an alternative to time-honored medicines, particularly in a sudden unanticipated crisis. It will, however, definitely help you relax in the ambulance on the way to the hospital.

Our body contains a strong yet not completely realized self-healing technique. Healing by crystals lets you access your already present healing capabilities. You should not experience any adverse side effects unless you buy a crystal thatyou can't afford.

Your innate skill would interest you in crystal healing, whether or not you have gems and stones that appeal to you. Don't inhibit your keenness, therefore, and go ahead.

Crystals are used to treat and strengthen the healing vitalities used. These magnificent stones may be used alone, in conjunction with the process of Reiki healing or other medical aid. They can be found all over the world.

The types range from various kinds of quartz to jade (a delicate gemstone requiring a high polish) and rubies (deep and dazzling red gemstones). Clear quartz crystal is the most widely used stone for healing and balance, but for various purposes, different types are used.

While talking about the working mechanism of crystals, these precious stones are used for safety, and to reinforce all remedial energy to its fullest potential. Each crystal has its unique qualities.

For example, transparent quartz balances and synchronizes your system and drives negativity away by creating positive energy within your system. Hematite is a building stone to discourage negativity and strengthen the body. Such energies happen by nature in crystals and ultimately will be reacted and controlled by the intrinsic force.

Your body can be healed in many ways using crystals. You can benefit from the healing properties of crystals simply by sticking them in your pocket or keeping them in your hands while you are sick.

Obtaining a Reiki form of crystal healing will bring the entire body together and stabilize it. The Reiki practitioners would place special crystals on the chakras (centers of spiritual power) of the body to push harmful energy into the body and instill curative divine and earthly energy. Many spiritual advocates and massage healers use crystals and can teach you how to use them by yourself.

Crystal healing is known to be a form of health care replacement. It is crucial to follow the recommendation of your doctor or physician. Nonetheless, you should research and gain knowledge of alternative therapeutic methods, such as crystal healing, and incorporate them into your life.

Crystal healing is an alternative type of treatment, using crystals and gemstones. The theory is that crystals also help to promote the healing process. This form of treatment has been used for headaches, insomnia, and cancer because recovery can be emotional and physical.

The crystals contain energy that can be transmitted to restore health and battle illness and provide spiritual guidance. Many proponents of crystal healing claim that while crystals do not explicitly cure disease, they can help solve the underlying problems. Every healer has his view, and some use the system along with other methods.

The fundamental premise behind the need for crystal healing is that disease takes place when the individual becomes unbalanced. Some healers believe this divine energy, or light, to be the very foundation of universal creation. It is believed that the crystals help to fix this disparity between us and our universe.

Some of the best crystal therapists were thought to have an ability to sense the exact location of energy in the body. This means that they can position the crystals on the body parts which promote healing most effectively. The stones of different sizes and colors promote a distinct healing quality.

Sapphires were said to contribute to mental stability, while amethyst stones were thought to encourage a calm mind. Rubies were said to help inspire bravery and purify the blood.

The crystals are all selected according to the needs of the person. It is not uncommon for some people to hold the crystals in their pockets and to put them into chains, to bathe with water, and to bring them around the room.

The crystal healing can be traced back to ancient times when Greeks and Indians thought of gemstones for another world. Many people believed that spirits existed in the crystals. Their use as an alternative therapy for some diseases is well-known.

CHAPTER 4. CRYSTALS AND THEIR METAPHYSICAL HEALING PROPERTIES

Crystals have long been known as beautiful and powerful artifacts. For thousands of years, people have used them for healing and protection. These strong gemstones were worn as a spiritual and temporal symbol of authorityby priests, rulers, and shamans. Egypt's leaders have been buried with gems that help their spirits enter the "other world."

In several respects, gemstones are valued. People sometimes wear a piece of jewelry with specific stones in it to preserve or improve their health. The therapeutic powers of natural gems and crystals are limitless and cures for endless symptoms of the human condition can be sought by understanding their use and application.

Throughout the ages, many cultures have used healing stones, so maybe there is some evidence. Stones, rocks, and crystals have the spiritual properties of healing. Such healing properties are seen as an effective complement to conventional medicine and healing methods.

Metaphysical healing is here and even associations with magic and occult. Perhaps this is why some people are vigilant about the whole thing and stay transparent.

There are many trigger points throughout our bodies. All these points, if stimulated by acupuncture or different forms of massage, create a healing effect. Stones are used to create the desired effect in combination with these trigger points.

In combination with their hot baths, even the Romans used stones to cure muscle aches and joint pains. Hot and cold therapy is an established therapy and so you can run straight out of the sauna in a cool bath or shower. Do this several times, and it is wonderful way to open and close your pores.

You can rub stones on your tummy to aid digestion or just roll a few stones in your hand for circulation and stress reduction. Abalone shell has electromagnetic properties which are ideal for protection of muscle and digestion, asthma and allergy. Many stones have purifying properties and can contribute to physical, mental, and emotional healing.

There is essentially a stone for increasing distress and you will find a listing of different stone properties in this chapter. The crystals are often used in the same way and have various forms of healing properties.

The very beauty of the crystals ensures that all kinds of jewelry are now made from them. In this way, you can pick the soothing crystal that you will wear as a necklace. This way the healing ability is still with you and not visible to anyone.

Many people also have healing stones and crystals in their homes, so each piece must be placed correctly to better impact the atmosphere of their home. Several crystal stores have emerged to satisfy the increasing swing towards crystal and stone healing.

If it looks great, you can just wear it or decorate your home and foster a sense of well-being in you.In this chapter, we'll take a look at the metaphysical properties of different crystals and how you can make the best use of them:

CARNELIAN

Carnelian is known as the Artist's Rock, which activates creativity, and potentially kills hidden talent and latent abilities. This is an outstanding gemstone for alleviating stress, relaxation, and overall well-being, this contributes to mental focus, increases physical strength and focus.

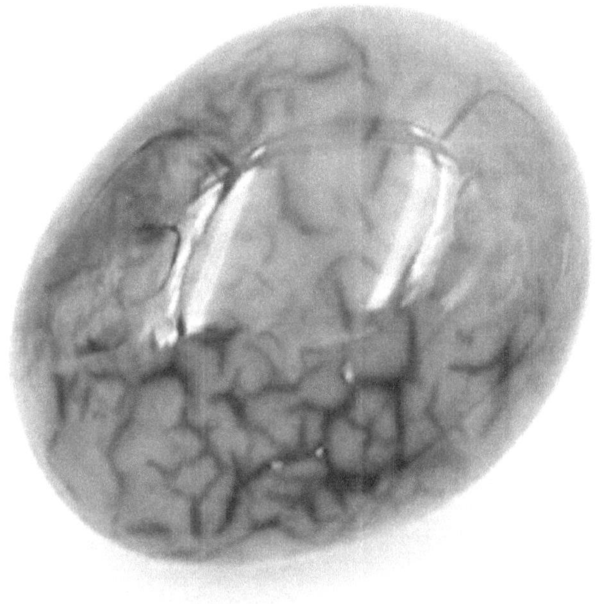

This resonates in the Red Crystal and Gemstones family with the Core or Important Chakra and the Sacred Chakra. With this beautiful stone, you can clean up and align your base chakra and feel relaxed and grounded.

Ancient Egyptians thought Carnelian was a source of power that maintained the cycle of rebirth and revival. The Egyptian Book of the Dead lists the use of Carnelian.

An exquisite material of stone was used by high priests to keep the vibration in their Chalice, as it has particular consistency. Most of the Chalices have silver coverings or designs, reflecting the cycles of moon rebirth and regeneration in the night sky.

Carnelian encourages a sense of confidence that allows users to better recognize the many choices ahead. Drinking carnelian loaded water will make someone with low self-esteem feel a sense of pride. This would also boost depression and empowerment.

- Healing properties: relieves stress, general well-being stone, increased concentration, increased physical strength, negative absorption, and repulsion.
- Vibration Frequency: high energy pulse, strength, stabilizing excess energy of the nerves.
- Spiritual properties: let the spirit float through the physical world, let it be in the actual moment.

As a personal favorite, this is one of the crystals and gemstone bracelets I wear every day.

RED GARNET

The Pomegranate Seed inspires the name of this precious stone. Like the other, his special energy isfiery red. In the past days, the Garnet was used to defend its wearers against hallucinations at night.

In answer to the second or sacral chakra, the gemstone aligns the wearer on various levels to attract wealth and prosperity, security, and strengthening relationships.

The vibratory energy frequency of this wonderful gemstone is that of the Third Eye, transparent, vivid, powerful, all-seeing, all-knowing.

The spectrum of colors is dark red transparent and nearly red and black. It is often offered as a sign of love and faithfulness.

While putting it on the Third Eye during meditation, it will purify negative energy, allowing you to see more clearly over time whether or not you are aware of the Third Eye opening.

It is strongly recommended to clean and charge the stone in the sun before putting it on your Third Eye Chakra. You can opt to move the stone through the burning or flame of a red or white candle rather than place it on the rays of the Sun. Whatever approach you use, always be vigilant when dealing with candles or incense.

DIVINITY ASSOCIATION WITH THE ANGELIC REALM

This gemstone is also associated with Angel Israfil, whose name means "The Burning One" who burns out imperfections, purifies love between two men.

- Healing properties: personal relationships, sense of security, enthusiasm, imagination, metabolism regulations, feelings of absence, fear of uncertain healing.
- Vibration frequency: powerful, penetrating, fiery purification.
- Spiritual Properties: Activation of the third eye, sympathy, vision reminder

If you would like to remember a past life experience, Garnet will help you fix what has already happened, taking you with compassion and affection to where you are now. Take a look before you head into the past; be sure if you want to revisit a circumstance or experience.

ROSE QUARTZ

Rose Quartz is commonly known as the "love crystal" belongs to the crystals and gemstones Quartz family. Rose Quartz varies from a light pink to a darker pink. It appears to shine and translucent in the sunlight. Often lighter rose quartz veins or striations appear almost white.

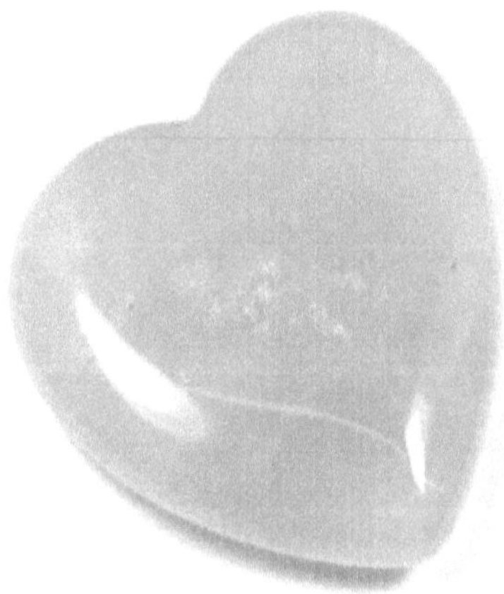

When held, you can begin to feel the harmony of your heart chakra in the center of your chest. Rose Quartz's gentle, subtle energy softens feelings and makes emotional wounds easier to heal.

Rose Quartz will strengthen the experience of meditation, relaxing you from inside. Place it on your table or pillow to help soothe your mind. A piece of rose quartz put directly on the chakra of the heart can gently create emotional wounds that will heal.

Emotional grief and suffering may be released by forgiveness. Love is gentle, soft and caring that allows you to heal at your pace. As Rose Quartz soft vibrations resonate with your Heart Core, it lovingly recreates the holy room inside, reconnecting you with Sacred Energy from the source.

In recovery, you should first give and receive affection. When you continue to curate, a void is created where pain and sorrow once were felt, which allows you to give and receive love, if you like.

When the energy of the soft Rose Quartz flows across your whole being, a sense of inner peace and harmony increases relaxation and healing. If you are looking for redemption with an open heart, it clears the way for new adventures and new experiences with a fresh sense of confidence.

DIVINITY ASSOCIATION WITH THE ANGELIC REALM

Archangel Jophiel whose name means "Beauty of God" gives one of us an option to see through caring eyes, embracing without a judgment what is before you.

- Healing properties: Emotional injuries, esteem, self-love, rising stresses, growing self-confidence, and fostering a feeling of inner peace.
- Vibration frequency: Medium, gentle, caring, calm, calming, subtle, vibration frequency
- Spiritual Properties: help you to align with your Self

SILICA QUARTZ CRYSTAL

Some of the theories of alternative medicine that have gained a lot of attention in recent days are that silica/quartz crystals have some healing properties and they can regain health for people who were previously sick.

This is a concept that some people in the conventional scientific establishment firmly reject and claim that it is unlikely for silica quartz crystals to heal and, on the one hand, we have traditional remedies who swear by the healing power of silica quartz crystals.

All this, then, leads to the question of whether the health benefits of silica are genuine.

There are now two ways in which we can address the question of whether the health benefits of silica crystals are genuine.

First of all, to find answers to the question of whether the health benefits of silica crystals are genuine is to understand how the silica crystals can cure us.

The effectiveness of silica crystals in restoring health is based on the belief that the loss of the balance of energy inside our bodies contributes to ill health.

As it turns out, our bodies consist of carbon, as is everything in the universe. This energy vibrates continuously. Or put simply, when the body vibrates at the right frequencies (i.e. when the body is in balance), good health is evident. Yet when the balance of energy in the body becomes skewed, ill-health presents itself.

To maintain good health, the lost energy balance needs to be restored and here the silica crystals are helpful, as it is observed that they have properties that are useful to restore it. When the energy balance is restored, health will return.

It is through this analysis of the process by which silica crystals cure that we can find a way to decide if the health benefits associated with them are genuine.

The other way to address the issue of whether the health benefits of silica crystals are genuine is to investigate the experiences of other people who have used them in pursuit of healing and this regard, we have a lot of people, some of whom are men of unquestionable honesty who swear by the use of silica crystals that their health was restored. This leads to the obvious inference that silica crystals may also be beneficial.

SELENITE

Selenite is a 'Protective Rock' of the Third Eye Chakra. You can use this crystal to protect against harmful forces and remove any negative energy. His use of force is for insight, psychic perception, comprehension, and also for communicating with the angels.

Selenite also ' Soothes & Calms,' which brings peace to you. Using little bits in the corners of your home to call for peace. Selenite also offers analytical insight and good decision-making skills, as it creates resolve uncertainty and allows you to see the bigger picture of a situation.

You can also use selenite to improve your memory. This crystal encourages sincerity and integrity, civil affairs, and connections.

It removes blocks of energy from physical and etheric bodies with selenite. This may also lead to removing electromagnetic radiation.

Please note that selenite is a soft and dry mineral.

Selenite uses:

You can use selenite for an emotional cure to encourage, harmony, forgiveness positivity, stress relief even to help build self-confidence.

Selenite is a marvelous mineral for a physical cure for backaches, bone problems, epilepsy, infections, insomnia, muscular problems, PMS, psoriasis, sleep problems, spinal cords, tumors, and ulcers.

Using selenite to connect with your higher self and/or upper worlds for spiritual healing, separate from negative forces, dissipate negativity, boost dreams, heal, meditate, encourage your past lives to remember, lift your vibrations, and work with angels.

AQUA AURA QUARTZ

Chakras' body's balancing is a vital aspect of being physically and emotionally powerful. Quartz is considered a particularly powerful stone for the creation and growth of the energies which flow through these important points. Aqua Aura can be used with a variety of crystals per every change you want to make.

Quartz can be found in many forms, but it is understood that it has some very special qualities. The hue is light blue, from an electric to a sky blue, with the iridescent shimmer of rainbow stripes and tiny points of gold. The metaphysical qualities of the stone are also more evident by the inherent strength of the stone and precious metal.

Blue quartz influences both the body and the spirit deeply. Nevertheless, the infusion of precious metals is accomplished by automatic methods, in which crystals are heated enough to condense and bind the vaporized gold permanently to the surface textures of stone. This transition results in a higher vibration that has a highly beneficial effect.

We can also amplify the vibrations of other crystals to achieve the desired effect.

Such crystals can be used in different ways to improve the patient's physical wellbeing, starting with improving the immune system and thymus gland. The use of this vibrating blood is often believed to aid with different forms of other disorders, such as autism, Asperger syndrome, polarity imbalances, dysfunction of the brain, and other genetic conditions. The results will be different in each situation.

This quartz greatly benefits from emotional well-being, particularly in meditation. The opportunity to make the whole person relax and control the energy flow helps people deal with stress, tension, and anxiety. The usage or holding of a stone removes harmful energy and makes it easier for you to be underwater.

This Blue Quartz will increase the energy of the soul and spiritual relationships during meditation. This can also develop analytical skills. Once you eliminate limitations, you can better offer latent gifts such as insight, channeling, intuition, and automatic writing. This also helps to protect against harmful electricity.

Quartz crystal communicates between the heart and the head, so you can speak truth and freedom from the heart. This particular stone can be linked to each of the seven chakras to heal each body's imbalances.

APOPHYLLITE

Most crystals and crystalline structures are structurally part of the Quartz family. Many types, colors, and sizes are available. Apophyllite is a rare crystalline tool for those who strive to improve their sense of consciousness, natural healing powers, and more simply connection to the Angelic Kingdom and the Kingdom of the Spirit.

The exquisite crystal forms natural pyramid shapes in its entire external and internal structure. While it is most frequently transparent, it often occurs in soft green tones. There is a shiny, subtle glow that first catches the eye and then the vibration tones start to fuse with your pulse, interacting on a deeper spiritual level. Apophyllite allows the user to relax the mind in learned ways during meditation. It is normal that a "tolerance" is progressively established for its particular vibration frequency, slowly or for short periods. Operating with this incredibly powerful crystal, you can experience vivid dreams, mystical hallucinations, and even higher levels of clairvoyance.

Apophyllite resonates with the Third Eye and the Crown Chakras, purifying it, clearing the internal pathways for an enlarged sense of awareness and improved spiritual development. Besides resonating with the Third Eye and Crown Chakras, green apophyllite also resounds with the Heart Chakra, stimulating and activated energy that matches the Universal Life and unconditional acceptance of the inner self and all that accompany you on your journey. It enables healers to adjust to their highest frequency and allows the purity of Divine White Light from the spirit world to flow into it and uses the human body as a healing energy channel through the user. All of Apophyllite's forms or colors will enable us to try inner healing on the physical, etheric, and metaphysical levels of consciousness.

Apophyllite is a very efficient energy transmitter due to its high-water content.

This so-called Stone of Truth helps the consumer to solve problems from the past, remove blockages or stagnating energies, and encourages improved metaphysical abilities.

DIVINITY ASSOCIATION WITH THE ANGELIC REALM

Archangel Taharial, whose name means "Purity of God" will enable anyone who can accept the special vibration of the Apophyllite into its own, incorporated and unified, to achieve a higher degree of vibration.

- Healing Properties: mental clarity; removes stagnant energies; helps eliminates toxic energy patterns
- Vibration frequency: very high vibration; solid white light transmitter.
- Spiritual properties: lift the veil between the physical and spiritual realms.

If you want to expand your natural gifts, intuition skills, heal your Spirit, work more closely with your Angels, or the Spirit Realm, this stone is a must-have.

JASPER

We must look at what crystal therapy is and what the features of jasper are. After exploring the Jasper estate, we will find how to use it for your advantage and the circumstances which can be improved by jasper.

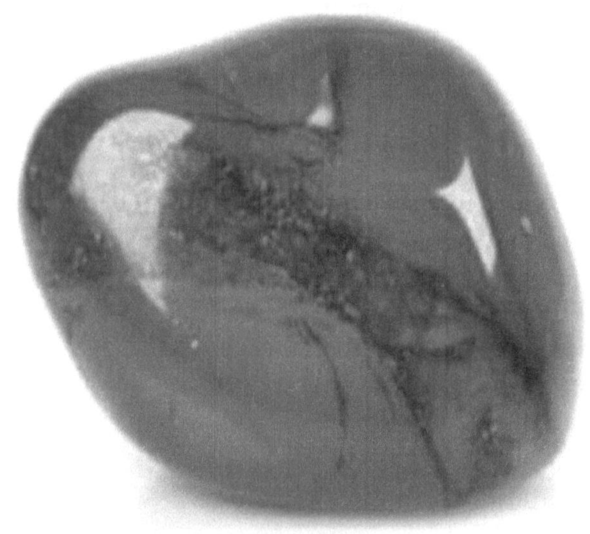

The idea behind crystal therapy is that every crystal has a certain amount of energy. This energy works on a person's body and mind and can contribute to a shift in mental and physical health. How crystals are used can be very important, but the basic concept is that an individual must regularly come near to a crystal. It is necessary to assess the situation thoroughly so that you can find the correct crystal for your particular case. Jasper is a quartz crystal that is often known as the highest nutrient. This is believed to be a stone that will help people in times of difficulty to bring peace and relaxation. There are many colors of jasper, and while it has general healing properties, each of the colors has its own.Some of the general healing properties of jasper include that it can lead to rest and absorb harmful energy. Jasper is thought to help eliminate environmental pollution, including radiation. It has many emotional advantages, including honest support and courage. Jasper users may find it to help them think rapidly and coordinate themselves as well as to inspire creativity. When you have a long-term illness, it will benefit you by healing your body.

I have also provided a list of properties of jasper types below:

1. Black Veined Jasper - this stone is a strong stone to stabilize and ground you. Some terms by which you can know it are spider web or zebra jasper.
2. Green Jasper - a stone of harmony that can help cure illnesses and obsessions. The stone is meant to increase the immune system. In particular, it is meant to help with skin conditions and help with bloating and digestive system problems.
3. Red jasper - this calming stone can lead to many problems such as anxiety and stress. This will help to fix unequal conditions and provide insight into challenging problems. This is intended to increase your dreams' potential. This may also help to detoxify the lungs, lungs, and liver.
4. Yellow Jasper – This is also a calming stone that can relax the nerves and emotions and is also useful for stress and anxiety. The digestive system is assumed to be helpful and help to reduce bloating.
5. Imagine Jasper - a stone of proportion and harmony that can add to your life's stability. This may also lead to the surface of suppressed emotions and help alleviate anxiety. This can activate the immune system physically and assist with skin and kidney problems.
6. Ocean Jasper - this stone will help a person to take responsibility and be patient. This may lead to improved digestion and eliminate body toxins that cause body smell.

AVENTURINE

The Heart Chakra resonates with Aventurine. The spectrum of colors is light to dark gray. It is typically found with smooth edges polished or tumbled. The stone is sometimes opaque, although the very light stones are transparent. There can be traces of mica in many Aventurine specimens that make them sparkle in the sunlight.

The frequency vibration of the green healing ray will help restore your eyesight and improve your emotions so that you can float through emotions of loss or uncertainty about finances or heartbreak more easily.

Green Aventurine soothes the emotions during times of stress and can protect the wearer from workplace disharmony.

DIVINITY ASSOCIATION WITH THE ANGELIC REALM

Archangel Jophiel, whose name means "Beauty of God" resonates with Rose Quartz, will allow you to heal emotional pain and frustration.

- Healing properties: peace, equilibrium, peaceful stressful circumstances or encounters, strengthen the dream, which is also called a full healing stone.
- Vibration frequency: safe, caring, calming, relaxing, gentle, constant.
- Spiritual Properties: remove emotional pain and mental pain; aligns inner vision to outer vision.

TOURMALINE

Tourmaline offers us a lot of green, purple, black, and blue colors. The crystalline structures are complex and reflect the nature of their vibration frequency on many levels. Clusters of smaller strands can be found that appear vertically connected in different sizes from small to large specimens about the size of your palm.

DIVINITY ASSOCIATION WITH THE ANGELIC REALM

Archangel Metatron whose name means "He Who Walks with God" allows you to balance your Chakra energy centers; healing sight, which opens your heart to love more completely.

- Healing Properties: bowel movement, nervous system balance, chronic fatigue relief
- Vibration frequency: churning, vivid, subtle, powerful, far-reaching, irradiating in every way;
- Spiritual properties: resonant with the highest and highest perception and healing subtle of divine growth

I was presented with a delicately packaged Green Tourmaline specimen with blue double endpoints. At the same time, it is strong and yet subtle. It helped me heal my Throat Chakra, aligning what resonates with loving kindness and compassion in my heart to express my truth.

You can wear it, place it under your pillow, place it on your napkin or throw a green light around you on your desk. No matter how you act with this green crystalline stone, be careful, do not force results. Experiences vary from person to person.

HERKIMER DIAMOND

These seemingly small crystals are very strong stones, which are of course double finished. Herkimer Diamonds are ideal stones for astral or out-of-body travel. This allows you to dream of all those crystals, minerals, and gemstones and amplifies them.

Externally, a Herkimer diamond, also known as the tuning stone can look too small to be successful.

Although it resonates with Chakra energy centers, it also resonates with the Crown Chakra more precisely. This stone, the Tune Stone, will allow you to balance and align all Chakra energy centers within your physical, emotional, mental, metaphysical, and Auric and Ethereal energy bodies.

DIVINITY ASSOCIATION WITH THE ANGELIC REALM

Archangel Taharial, whose name means "Pureness of God" will allow you to purify your emotions. Archangel Ariel, whose name is God's Lion, will help you get your relationships into order.

Archangel Haniel, whose name is Glory of God's grace, will help you to grow and extend your insight and inner vision.

Arcangel Elemiah, Angel of the Sea, will help you to heal emotional disharmony and to allow you to move inside.

Mystical Goddess, Isis will allow you to access your inner guidance and increase your sense of consciousness.

Ascended Master, Lord Krishna will help you rediscover love, harmony, and happiness, radiating goodwill to all those who pass on your path.

- Healing properties: multi-dimensional recovery, stress relief, and concern,
- Vibration Frequency: very strong, amplifies all other crystals' energies
- Spiritual Properties: telepathy, regeneration of the spirit, combining physical and etheric energy bodies.

Herkimer diamonds are powerful transmitters so they can help you with astral journeys, deepen your meditation experiences, and explore your Dreamtime. You will also increase the sense of consciousness and telepathy among two people who are closely related.

Herkimer's energy is very powerful, but at first, it will seem subtle before you start interacting with it every day. Be open to its many attributes and you will be surprised at your next trip.

KYANITE

Blue Kyanite is one of the few crystalline forms that can renew its energy frequency. Unlike most other crystals and gemstones, it does not require cleaning. You can express your gratitude by periodically putting this beautiful crystal by the sun or moon. It strengthens your link to this blue crystal.

This special crystal is capable of blocking all contact channels and enhancing your conscious connection with the World of God, Angels, and Spirit Guides. It is suggested that you start to work with this pillar during meditation to match your energy frequency with its harmonious vibration, which combines light and darkness.

When negative energy is cleansed and dissolved, your inner healing process will be accelerated and you can free yourself from your past. There are many ways, many channels of communication that are opened physically and spiritually during the healing process.

With both the Throat Chakra and the Third Eye, your energy center resonates for vision both internally and externally. Through using positive thinking strength, aligning your inner self with your divine energy system, you can reach your inner vibratory frequencies, tuning the angelic domain and the spirit world.

DIVINITY ASSOCIATION WITH THE ANGELIC REALM

Archangel Michael, whose name means "He Who's like God" can purify from inside and around you less toxic energy, when asked.

Archangel Gabriel, the Messenger of God, will allow you to remove blockages in your Throat Chakra so that you can express your truth with clarity and comprehension.

- Healing properties: dispels toxic energy, activates the immune system, clears the path ahead of you, and also balances chakras
- Vibration frequency: quiet, free, higher consciousness
- Spiritual properties: Spiritual world contact, dream recollection, telepathy

If you like, you can fill the glass with filtered water and put kyanite into a glass to allow it to charge before you drink. You can also bring multiple bits of kyanite into your pitcher of the tap water filter. The efficiency of this system depends on how big your pitcher reservoir is, how much the kyanite is filled, and how many parts you have put on the surface.

As always, the crystal vibration frequency can be increased by adding transparent quartz.

AMETRINE

A stunning natural mixture of amethyst and citrine quartz helps someone who wants to identify closely with the Spark of Divinity. Ametrine promotes self-expression, acceptance and increased self-esteem, and dissolves barriers in the emotional energy body and the field of Auric energy.

This crystal will help to refresh the etheric body, bind the user to the spirit world, and reach ever-higher consciousness levels. The crystal pair covers the individual whether the body is damaged or used. It encourages the transfer of lower energy vibrations and provides a strong energy cleanser to help the physical body and aura transmute harmful energies.

Ametrine resonates with Solar Plexus, Third Eye and Crown Chakras, which create a special energy signature for all those drawn into the crystalline high energy shape. When amethyst and citrine come together, the greenish-gold color is always faint. The crystal's vibration is incredible!

The Solar Plexus is the object of mental insight, expression, imagination, and empowerment.

The Crown Chakra is the actual entrance into our relationship with the Divine Energy Source. Ametrine helps the solar plexus and the power core Crown Chakra to merge their particular energies into one another, integrating the metaphysical and physical realms.

The Third Eye chakra is the source of inner vision, divination, insight, and clear vision.

DIVINITY ASSOCIATION WITH THE ANGELIC REALM

Archangel Michael, whose name implies – Who is Like GOD will help you overcome uncertainty, which also includes lower negative thoughts emotions, and thoughts.

The Archangel Sandalphon, whose name implies God's strength and glory will help you concentrate your thoughts and understanding.

Ascended Master:, ST Germain would enable those who try to free themselves from the limits of their convictions to free themselves and to encourage others to grow higher to realize the Spark of Divinity within.

Merlin, the divine wizard, allows us to reconcile the truth of what remains in physical form with their inner spirit.

The Ascended Master of Beauty, Knowledge, and Intellect, Lord Kuthumi. If asked, he will support and inspire you to discover or rediscover self-love. Self-love opens the road to everything you desire.

- Healing properties: realization of self, breakdown of emotional barriers, defense, self-consciousness, manifestation.

- Vibration Frequency: efficient energy purification, physical and Aura detoxification
- Spiritual properties: insight, discernment, the relation to a spiritual world.

As you become gradually aware of who you become, you are more conscious or mindful of your physical life and spiritual experience as it unfolds before you.Keep Ametrine in your left hand as you meditate, it will improve your experience.

CHRYSOCOLLA

Blue symbolizes harmony, serenity, emotions, and infinite capacity. The Peace Stone, called Chrysocolla, has a very high frequency of electricity. Gem Silica is another name for this amazing healing stone.

This crystal is renowned for its amazing virtues of harmony and justice, the user should take a deep breath of emotional suffering and sorrow. As a healing rock, it is possible to cure both the Heart Chakra and the Throat Chakra at the cellular level by removing stones.

Gem Silica resonates with the Throat Chakra and dissolves any fear of saying your truth, knowing your worth. When you speak the truth, remember to talk in your voice, energy, and eyes with love from the heart.

Chrysocolla works with The Heart Chakra to balance emotions, heal from past and present wounds, and to instill a feeling of peace. You can hold Chrysocolla in your pocket, worn as a pendant between the Throat and Heart power stations, in your home at your office. When you place it under your bedside table, it will work for you while you sleep. Emotional healing can be subtle or dramatic, creating an energy shift that changes your life.

DIVINITY ASSOCIATION WITH THE ANGELIC REALM

Archangel Raphael whose name is "He Who Heals," will help you to heal your spirit. Archangel Gabriel, whose name means Messengers of God, will assist you with loving-kindness and honesty in expressing your facts. Angel: Kokabiel, Angel of Stars, will help you to keep your eyes on the road ahead of you and light up every step of your journey.

Anpiel, Angel of the Sky, will allow you to see your infinite ability beyond physical limitations.

Ascended Masters, El Morya Khan, will help you gain an insight into all your life's goals and beyond with ancients' spiritual truths. Master Hilarion will help you with the courage to create your reality, heal your self-worth sensation, and open the door before you to unlimited possibilities.

- Healing Properties: heartbreak emotional wounds, blockages of the throat chakra, soothe broken nerves, calms emotions; endurance, compassion; tolerance; endurance
- Vibration Frequency: Exceptionally high vibration; male and female energy balances
- Spiritual Properties: increase your connection with the Divine Source, reach through a veil between the spiritual realm and the physical domain of matter

Chrysocolla can bring peace of mind and heart to the user during meditation. Keep this beautiful healing stone in your left hand, let the vibration flow through you and purify, balance, and align your spiritual, physical, and emotional energies with the Divine Intent for your life.

SMOKY QUARTZ

Smoky Quartz is a dark joy stone that can help protect you from psychological assault, the projection of other negative thoughts and emotions. One soothing product is to help you remove toxic energies from your own heart and mind.

It has the power to canalize pure Divine Light into your Root Chakra, which will induce harmony and transformation in the entire physical body and energy bodies.

Smoky Quartz resonates mainly with the Root Chakra or Essential Chakra at the base of your spine. It will allow you to focus your thoughts and emotions. This will also motivate you to accept the difficulties you face throughout the way and take responsibility for your way.

Smoky Quartz also resonates with the heart center to help alleviate stress and heal emotional injuries through the release of harmful patterns of thought. The dark quartz can draw energy from the Heart Chakra to the Root Charka, making it easier for you to work with your emotions.

DIVINITY ASSOCIATION WITH THE ANGELIC REALM

Archangel Raziel, whose name implies the Secret of God, upon request, will help you shield your minds and feelings from the destructive thoughts of others.

Yabbashael, Earth Angel, will help you ground excess stress and negative energy.

Tara, mysterious goddess of light and love, will allow you to see the light even during dark times. She extends or assign love without prejudice to you and others.

- Healing properties: emotional exhaustion, stress, over-energy lands, concentration feelings, psychic attack shield.
- Vibration Frequency: solid, calming, and converting negative excess energy to transmutation into the light.
- Spiritual Properties: assimilates awareness and insight, dissolves negativity.

When you are depressed, please call Archangel Raziel, Angel Yabbashael, or Mystical Goddess Tara to help protect you, calm you down, and ground you.

It can be worn on the body, put in your pocket, or put on the bedside table. It can also be put on your desk in your office to help create a barrier or filter to prevent harmful effects on the workplace.

BLACK TOURMALINE

Tourmaline is a special crystalline form, opaque and translucent at times. You can know Tourmaline's many colors: blue, green, pink, and black. It is previously known as Schorl, Black Tourmaline.

Known as the calming stone, it can absorb and help you remove the negative energy around you, allowing you to dissolve inner, mental, and emotional blocks of energy.

There are so many changes, so many transitions around us that it is important to protect you and your Auric energy field so that you can maintain your concentration and balance. It is essential to maintain balance to repel negativity.

Black Tourmaline mainly resonates with the Root Chakra at the base of the spine. The Root Chakra Energy Center helps to draw energy to your physical body from Earth's Mother, balance male and female energies, and detoxify your Chakra Energy Centers.

This provides a shield, a defensive energy shield that can help you repel harmful charged energy that neutralizes the effects of the psychic attack. Psychic attack is described as negative thoughts like jealousy, wrath, and hate directed from others to you and yourself.

DIVINITY ASSOCIATION WITH THE ANGELIC REALM

Archangel Michael, whose name means He Who Is Like God, will help clear you from and in your environment of all negative and dirty force.

Ascended Master, Lord Kuthumi will enable you to be wise, caring, and thorough. Learning how to expel anxiety from inside protects you from the harmful energy of others.

- Healing Properties: calming excess nervous energy, detox fluid, energy balances in and around you.
- Vibratory frequency: intense energy level that incorporates the energy of Mother Earth into the human body to achieve equilibrium
- Spiritual properties: fosters confidence and strength; motivates spiritual energy

This stone will inspire you to have confidence in times of stress, difficult circumstances, or encounters, and find strength internally to release the fear of the unknown, doubt, and anger.

Place Black Tourmaline on your desk, at your bedside, or take it with you at the front door of your house. When you become aware of who you are and of what you want, during your transformation Black Tourmaline will become a trustworthy friend and partner until you achieve equilibrium within your energy centers and mental, emotional, physical, and spiritual energy bodies.

TANGERINE QUARTZ

Also known as the Soul Stone, it provides security from other people's psychic attacks. This will also defend you from harmful self-sabotaging feelings.

If you want to increase the transition from the spiritual realm into the physical form, work with Tangerine Quartz in meditation, or use it as a method to focus your thoughts. This boosts the flow of imagination and enables you to release the self-made restrictive convictions of modesty you want and more.

Tangerine Quartz resonates with the Solar Plexus Chakra with the central energies of autonomy, self-esteem, discernment, and manifestation. The Holy Chakra often resonates with an earthy-orange dark color. The Holy Chakra is the source of energy for mental and physical relations.

DIVINITY ASSOCIATION WITH THE ANGELIC REALM

Archangel Uriel, the name of whom means the fire of God, will allow you to manifest all that you wish.

Ascended Master Ganesh, if asked, will help you eliminate obstacles from your path, direct your inner wisdom, and aid you in manifesting wealth and abundance.

- Healing properties: multi-dimensional healer, liberating trust, creativity improvement, self-esteem problems, inner self-empowerment
- Vibration Frequency: energy, power, feverishness, warmth, comfort, encouragement
- Spiritual Properties: strengthen natural intuitive abilities and express and align the Auric energy body;

This has the power to eliminate the shadows and enhance your inner vision to allow clarity and discernment on your journey.

The Soul Stone is beautiful because it weaves its special vibration frequency within and around you, shielding you from the harmful energy thoughts of others.Past-life problems will emerge when you widen your knowledge. This orange crystal tends to weaken you in curing issues from days gone by.

You will find out how to disintegrate self-sabotage convictions, self-creating barriers, freeing you to represent your true existence, your dignity of mind, and spirit.

NIRVANA QUARTZ

Nirvana Quartz, also known as Ice Quartz, was first found in 2006 at elevations 18,000 feet above the sea level in the Himalayan Mountains of India. Melting glaciers have exposed this amazing mineral gift to all of us with a chance to witness remarkable shifts in our levels of energy. It is shaped irregularly; no two mineral specimens have the same shape, color, or frequency.

Why can each mineral specimen have a particular frequency of energy?

Just as every bird, tree, and human has its unique energy signature, every ice quartz has its energy signature. Each specimen also shows jagged edges and gaps. The color range is almost translucent white to soft rosy pink shades. Nirvanais the freedom of the soul from the burdens of karma and physical life according to Hinduism beliefs. Nirvana is also a feeling of putting everything behind you now.

This is the freedom to liberate the conscious mind psychologically from the limitations of being confined and it permits the thoughts and spirit to swell among the stars of the heavens above. The Word Nirvana means also a state of light, a sense of clarity and more open communication with the Divine

Energy Source. It is a passage for deeper stages of mystic lighting, transformation, and transcendence. This also means a feeling of complete serenity, peace, and satisfaction.

It resonates with the energy centers of the Third Eye, Crown, Throat and Heart Chakra and incorporates each other. Use Ice Quartz meditation to increase your awareness and expand your inner essence and spirituality. This crystalline form will allow you to discover and understand the meaning of your life. This will help you to relax your mind, interrupt or quiet your mental speaking so that you can be guided.

DIVINITY ASSOCIATION WITH THE ANGELIC REALM

Archangel Jophiel, whose name means Beauty of God, is going to help you to conquer emotional pain and sorrow that allows you to transcend time and space and to heal deeply in all aspects of time and space.

- Healing properties: dissolves internal emotional wounds, facilitates self-acceptance, deepens meditation.
- Vibration Frequency: high energy transmission, subtle in its journey from Spirit to physical matter.
- Spiritual Properties: the meaning of creation, astral projection, dissolves boundaries from the physical world to the spiritual realm.

Nirvana Quartz will build in you a sense of growing consciousness, deepening experiences of meditation, and promoting self-acceptance. This is the Ethereal Realm's record keeper, it accesses astral visions, helps remove barriers, and clears the way between the physical world and the Angelic Realm.

SUGILITE

Violet is the color most commonly associated with love and our link with Divine Power. Sugilite is among the other stones known as the Stone of the Self-Forgiving,Unconditional Love. This color breaks down all the negative energies in and around us.

Sugilite stands for the powerful connection between the mind and the physical body whose purpose is to control our mental processes consciously.

We can manage our thought processes without manipulating them, our mental capacities allow the healing force of the Cosmos to flow through our physical bodies, which balance the physical with the etheric energy.

This is a very strong stone or mineral that Earth Mother has given to us since the early 1980s. This can be put on the Third Eye Chakra to cleanse, soothe, and balance your inner vision and match it with your outer vision. This helps you to see beyond physical boundaries.

The Higher Consciousness Stone coincides with the Crown Chakra, Third Eye, and Eighth Chakra found on top of the Crown. Working with Luvulite or Sugilite as it is also called, a medium is generated to obtain vast quantities of healing energy that radiates in all senses of time and space on all levels at the same time.

To the empathic healer and for those who are extremely intuitive, it will help you balance your feelings and all that you experience and put you back into harmony with your true self.

Luvulite bears signs of tangible matter from the angels to the earthly world. This will help you to reconcile what awaits you with all that you left behind. The high-energy mineral stone eliminates doubts, which allow the user to transmit and to radiate the energy frequency from the violet-red light.

DIVINITY ASSOCIATION WITH THE ANGELIC REALM

Vihianna, The Angel of the Ray of Purples, is helping all those who ask to improve the conscious consciousness and spiritual growth with Crown Chakra.

- Healing properties: frees anger from past times; dissolves frustrations; strengthens self-confidence that softens feelings of fear and distress.
- Vibration Frequency: very strong, efficient, low energy vibrations disintegrate
- Spiritual properties: spiritual expansion, awareness-raising, meditation deepening.

I can see and hear my physical body releasing the stress for me to center all my energy bodies and to re-align them with Divine Energy sources.

TANZAN AURA QUARTZ

This rare crystal is also called Tanzine or Indigo Quartz, which vibrates on a far higher level of consciousness. This facilitates contact with the Heart, the Higher Awareness, and strengthens the Ethereal Realm awareness.

Tanzan Aura Draws divine energies to the physical body and canals very powerful vibrations through everyone interested in this magical quartz crystal. It helps the consumer to funnel excess energy into the Earth Mother to be converted into the light of the Divine Source.

The more you work with Tanzan aura quartz, the more it allows you to make your dreams transparent. It will help you with a dream interpretation that allows you to connect with inner self at deeper levels than you could have before you started working with this crystal.

The Indigo Quartz resonates with the energy centers of Throat chakra, Crown Chakra, and Third Eye Chakra. Its main energy source in the universe directly influences or communicates with the Soul Star, the Chakra energy core just above the Crown Chakra.

Tanzan helps to improve intuitive skills, to improve and foster trust in received messages. This also provides multi-dimensional equilibrium that helps the individual to reconcile the physical world and the metaphysical domain, thereby extending the conscious mind to see beyond the physical world.

DIVINITY ASSOCIATION WITH THE ANGELIC REALM

Archangel Ophaniel will help you restore your inner equilibrium with what is happening in and around you.

- Healing Properties: reduces stress, restores equilibrium on all levels at the same time, restores inner vision, confidence.
- Vibration Frequency: very strong, the celestial force pulls into the extremely mystical physical domain
- Spiritual properties: Improved perception, multi-dimensional harmony, increased spiritual awareness.

Tanzan Aura Quartz is transparent quartz bonded to the rare mineral Indium, the gold, and the niobium that creates a blue-violet color with a highly subtle energy level. Like all blue ray stones and crystals, it allows you to communicate, express your heart, relax your feelings, instill a sense of peace and harmony within your inner core of reality and beauty, love, and light. It will allow you to open your Third Eye and restore vision as you search for answers to the meaning of your life, gain clarity, and the confidence to continue on your path. If you want to "explore" other worlds and realms, the Tanzan Aura Quartz is your path to astral projection through the many Milky Way corridors and other Stern Galaxies.

IMPERIAL GOLD AURA QUARTZ

This is also a stone of wealth and abundance that allows what you want to manifest into physical form. It has the Sun color which depicts energy, self-improvement, regeneration process, and strength.

Imperial Gold connects the Solar Plexus Chakra with the heart chakra and bridges the high Chakras of the Soul with the physical or lower Chakras of the earth. This crystal will allow you to concentrate your thoughts, to manifest the wishes of your heart; to give comfort and emotional balance.

If you want to use Imperial Gold, allow its powerful vibration, beginning with your essence, to transfer your focus. Be open to modifying your inner thought habits, emotional reactions to current circumstances or experiences. Willing to release energy blockages that trigger unconscious self-sabotage attitudes and build distance between you and everything you want.

Imperial Gold is a strong contact stone, with the capacity to extend your sense of awareness and understanding. This will encourage you to open your Third Eye, so that you can see your life as it is as you are giving it to see what is possible, but allow it to unfold before you.

Be willing to use this quartz crystal in meditation to integrate excess energy between mind and body, calm and grounding. This encourages confidence, clarity of mind, increases resilience, and inner power. This also allows you to concentrate on the future instead of just looking at the past.

DIVINITY ASSOCIATION WITH THE ANGELIC REALM

Archangel Uriel, whose name means the Flames of God, will allow you to manifest all that you want in your life. He will allow you to build your truth rather than the truth that others would have for you.

- Healing properties: mental equilibrium; Solar Plexus Chakra energy center equilibrium, the physical body core; contact between Heart and solar plexus chakras. It also transmutes toxic energy.
- Vibration Frequency: strong power source that radiates and projects the energy of the Sun and that magnetizes you.
- Spiritual properties: concentrate your intentions; manifestation; mental awareness; merge the physical world with the spiritual realm.

Remember that everything you want to have, to be, and to do with your life is already present in the Realm of Spirit and you just need to be able to embrace everything you have asked for.

ANGEL AURA QUARTZ

Angels are everywhere around us in every situation and experience. We have to ask Angels to support us; they look forward to your calling. No request is too large or too small. Quartz Angel Aura is shimming in the light. His subtle vibration is very powerful, enabling us to enter the Kingdom of Heaven, the Kingdom of Spirit, through the veil of forgetfulness.

No request is essential, these are concepts of the physical domain. If you have requested support, be ready to obtain what you have requested. In case of doubt, request confirmation or clarification of what you got.

When you chose to work with this special quartz crystal, it will increase your experience of meditation and strengthen your bond with the Angels. It protects you both inside and around you from unnecessary harmful energy. Holding, transporting, or wearing Angel Aura Quartz instills a sense of calm and helps you to grow a sense of harmony that flows deep inside your true self.

The Heart Chakra always resonates because Angels represent the unconditional love and complete acceptance of all life and everything.

The beauty of this quartz crystal lies in the association of exquisite, transparent quartz with platinum and silver, which produces a shimmering look identical to the wings of Angel. This brilliant surface often reminds me of the Moon when it is filled up in the sky.

DIVINITY ASSOCIATION WITH THE ANGELIC REALM

Aurora Angel of Crystals will aid you in connection with Divine Origin Light, Realm of Angels, Realm of Spirit, and your inner soul if requested.

- Healing properties: Restoring the Spirit by the core of your body and mind, fluidity, and harmony.
- Vibration Frequency: very high, subtle, soft, and strong.
- Spiritual Properties: strengthened the connection to Angels, Crown Chakra extended, 8th Chakra opened

Angel Aura, as its name suggests, stimulates consciousness of our inner light source and the Divine World full of responsive Light and Love beings, conscious knowledge. There are angels in many forms, human, ethereal, and the wonder of nature around us. Use this improved crystalline diamond to match the physical and etheric Chakras. When you want to work with this amazing stone, be open to your own special experiences.

WATERMELON TOURMALINE

Many of us wonder whether we'll ever find our Soul Mates in our lifetime. For those that have and are with their Soul Mates, questions arise about "now what" or "is there something more I am supposed to be doing with my life?"

Watermelon Tourmaline is going to assist anyone who is searching for their Soul Mate. This also allows both matched Soul Mates to have insight, emotional remedy from past problems, and comprehension at a soul level with greater clarity.

The Watermelon Tourmaline blends purple, blue, and pink tourmaline in a beautiful and unique crystal. The green color can be variable from light to dark. This is a stone or crystal with incredibly high vibrations.

This incredible crystalline stone will reassure anyone who wants to understandand affirm that they truly fulfill the purpose of their life. This will also allow you to gain a better understanding of the meaning of your life. Meditating on this powerful crystal will help you unlock the hidden knowledge inside.

Multi-color Tourmaline resonates with more than one energy center, integrating the Heart and Throat Chakras, and then all other Charkas. Other varieties are listed below:

1. Yellow Tourmaline: mainly resonates with the Heart Chakra. It will allow you to balance the entire physical body. This can align all the energy centers of Chakra at the same time. It is a very strong vibrant stone that you should appreciate and work with slowly to prevent your physical body from being over-stimulated.
2. Green Tourmaline will help you with spiritual development, in a subtle yet deeply understanding way.
3. Pink Tourmaline works with you to loosen heart blocks and to encourage love to flow through you, to give all those that cross your path a beautiful and delicate glow. Pink primarily resonates with the Chakra of the Heart, affection, and emotional healing on all levels, which include cellular levels in your physical and etheric energies.
4. Blue Tourmaline: mainly resonates with the Throat Chakra, allowing you to express your truth with respect, compassion, and clarity. Speaking your truth without evil and judgment is one of many ways to open the key to your higher self and to align yourself more directly with the spiritual realm.

DIVINITY ASSOCIATION WITH THE ANGELIC REALM

archangelAurora, Angel of Crystals is welcomed to help those who pursue emotional healing on all levels, beginning with the heart and throat, which permeate all Chakra cellular energy centers. Ask Aurora to help you discover or better understand the meaning of your life.

- Healing properties: balances past emotional trauma, dissolves mental blockages; chronic fatigue; health and wealth issues; removes Throat Chakra barriers;
- Vibration Frequency: very high frequency; very strong without imagination.
- Spiritual Properties: Spiritual Expansion; improve your connection with your Higher Self, increase conscious awareness, disclose the meaning of life

Tourmaline is a rare gift to anyone who loves this magic stone. If Watermelon Tourmaline has endowed or found you, embrace its many secrets, gifts of love, success, and wealth, and discover the meaning of your lives. Be ready to trust Tourmaline's many opportunities in any shape or color combination.

FLUORITE

A rainbow can be one of the most impressive light shows alternating seamlessly through them. This is a sign of hope, harmony, and regeneration of water. A bridge symbolizes a path from one horizon to another, a connection, and a transformation.

Fluorite, also known as the Rainbow Bridge Block, includes all the Rainbow colors. Tumbling specimens make the observation of the many colors simpler.

For example, the green-ray vibrates mainly with the heart chakra, wellbeing, and abundance as it incorporates all of the colors of the rainbow within its internal structure.

The purple light symbolizes spirituality, wisdom, and improved consciousness that mainly resonates with the Third Eye and Crown chakras and also contains all the rainbow energy levels,. The blue-green color binds and incorporates the chakras of the Heart and Throat.

Most specimens of fluorite show white light, pureness, and dignity. White light symbolizes our link to the Divine Power Source and the Angels who direct and protect us along our respective paths.

Fluorite varies from a soft, almost pastel violet to a very dark, deep vivid purple, green and white colors. Do not be misled by your poor eyesight, these exquisite specimens beautifully illumine all the colors of the Rainbow.

I recall when I first saw and held in my hand this beautiful crystalline stone structure. It felt as though its particular frequency of energy, coupled with mine, interwoven inside and around me. I just have to look at Fluorite to feel this feeling over and over again.

DIVINITY ASSOCIATION WITH THE ANGELIC REALM

Archangel Taharial, whose name is "Purity of God," is an ultimate guardian of the Celestial Rainbow Bridge and the Angels of Rainbow Bridge. They will direct anyone who asks for his help to discover their special path into themselves during the journey of existence.

- Healing properties: Third Eye and Crown Chakras cleaning and balance; emotional wounds, Chakra energy centers combine, incorporate emotional, physical, mental, spiritual, and ethereal energies; release excessive excess energy.
- Vibration Frequency: pure white light, rainbow light prism, subtle, powerful, soothing, calming, unifying.
- Spiritual Properties: Divine Source relation, integrity, insight, conscious consciousness extension, integration.

Fluorite has the power to simultaneously clean, align, and incorporate all your energy bodies and all your Chakra energy centers. This special stone is matched to your current level of vibrations. When you expand your knowledge, Fluorite can align itself to your advantage.

CHALCOPYRITE

Peacock Rock, also known as sunlight chalcopyrite. The colors of each mineral specimen differ. Pinks, blacks, greens, and variations produce delightful images! Chalcopyrite contains sparks of a gold-like substance, which is pyrite, most frequently. This is known as the stone of Chakra.

As the Chakra stone, all the energy centers are washed, balanced, and aligned individually. When everyone is fully aligned, pure, and healthy, this beautiful multicolored stone will align the physical body at the same time with the metaphysical and etheric energies.

It will help you change your inner self, balance your thoughts and emotions in a very soothing and relaxing way while working with Chalcopyrite. Furthermore, Peacock Stone activates the Third Eye, so that the consumer re-discovers long-forgotten information.

This is the secret to reconnecting with old mystical experience. Hold it over your Third Eye or in front of it during meditation. It will help you to awaken your inner sight in time depending on how open you are.

One of Peacock Rock's many benefits is the production of excess nervous energy, which can alleviate stress-related emotions and then create an equilibrium in the body.

DIVINITY ASSOCIATION WITH THE ANGELIC REALM

The ArchangelMetatronwhose name means, He Who walks with Heaven, will allow your whole Being to bring your energy centers into harmony within your physical, emotional, mental, spiritual, and Aurical energy fields.

- Healing properties: lines, cleanses, balances all Chakra Energy Centers; soothing fears, controlling lack of thoughts; eliminating energy blocks; increasing self-esteem.
- Vibration frequency: tuned to or with all of the universe's vibratory frequencies; gentle, heavy, subtle, and efficient.
- Spiritual Properties: psychological protection, healing, insight, and vision improvements.

Chalcopyritecomprises of the Copper part and Pyrite part.

Pyrite is also referred to as "Fool's Gold"; its spiritual properties are prosperity and abundance, displaying great possessions, and eliminating what is no longer needed or desired. Copper's spiritual properties resonate with the Solar Plexus Chakra, a source of energy for creation, and are hot and healing.

Unlike other color stones and mineral specimens, the interconnection of each part produces a unique vibration of energy and color. That stone shows a stunning variety of colors that resemble a male Peacock's tail feathers in all their beauty.

ROSE CALCITE

The self-healer inside us resonates with the rosy light, the gentle and subtle, yet deeply penetrating rosy, caring energy. The central aspect of spiritual growth is redemption. To learn to forgive oneself and others is to learn to pardon oneself and others for past errors or mistakes.

Cobalt calcite is often generally referred to as cobalt, rose, or rose calcite. The color ranges from light rose to dark rose. The crystalline structure, the feature, when it glows in the sunlight, is remarkable. The penetrating energy of this crystalline shape draws you to its healing rays of unconditional love, which open your Heart Chakra.

Rose Calcite resonates and communicates with the energy centers of the Heart, Crown, Root, or Essential and Sacred Chakra. The Third Eye is allowed to see deep-rooted emotional wounds in the physical world, lovingly removing the thing behind you.

The Chakra of the Heart and Crown integrates, establishing direct contact with the God within, enabling a deeper sense of universal love and a sense of happiness.

Although the crystals are small, do not underestimate their capacity to heal emotional injuries from the past if you are willing. It is a healerstone unlike any other.

Distance healing involves more than merely transmitting light and energy through the ethers of space and time in the physical world. Distance healing enables the physical mind to "step aside" so that universal love can be performed at the cellular level from the divine source of everything.

DIVINITY ASSOCIATION WITH THE ANGELIC REALM

Archangel Jophiel whose name implies Beauty of God will help you in healing emotional sores and in seeing beauty in everything, in every facet, everywhere. Archangel Raphael, whose name is He Who Heals, will help you to serve yourself and others.

- Healing Properties: emotional wound, sympathy, heartbreak, repentance, the release of emotional obstructions, removal of defensive emotional obstacles
- Vibration Frequency: Sweet, caring, universal, subtle, gentle, compassionate
- Spiritual properties: Eternal Presence, transmutes traumatic experiences into the sun

Use this mineral to open your heart chakra during meditation and to expand your radiant loving power. Enable yourself to be a beacon of caring rose light that inspires people to find their way into a sea of despair, a world without love.

Be open to this beautiful mineral's loving strength. Its elegance is unmatched. This will strengthen your relationship with the Spirit world in ways that you still have to discover.

PINK TOURMALINE

Rubellite, also called pink tourmaline, is well known for its soft but strong pink vibration frequency. This is one of three Trinity Rocks, each linking spirit, body, and mind. This varies in color from pink light to dark rosy-red.

Emotions generate thought. Thoughts generate intense feelings. To merge and balance your entire being, using Trinity Rocks, Rose Quartz, Kunzite, and Pink Tourmaline, combining the energies individually and eventually from within to establish synchronized harmony. Pink tourmaline interacts with and retains the lower physical body energy centers of the higher spiritual Chakra energy centers. It will help you reconcile male and female strength, the yin yang pull of duality into the physical world. This strengthens contact between spiritual and conscious mind worlds, ignites the direction, psychic knowledge that has been provided intuitively, and extends the sense of mysticism. Rubellite can be used to restore ties between friends, family, and lovers.

Do you like to unlock the past? The crystalline stone will reach deep into your emotional energy structure, purifying, and healing all the ways of time and space if you are willing to be fully honest. You can free from the resurrection of old emotional tapes that trigger additional pain and sorrow.

During meditation, put Pink Tourmaline during your Heart Chakra to encourage emotional healing and to dissolve the superficial layers of defenses which are no longer useful for you. It can also be used as a pendant near the energy center of the heart.

DIVINITY ASSOCIATION WITH THE ANGELIC REALM Archangel Ariel whose name means, Lion of God, will help you heal emotional injuries, awaken trust, inspire faithfulness.

- Healing properties: heartburn, balance male and female energy, eliminates blockages of energy from the emotional energy body
- Vibration Frequency: powerful, divine love for healing, calming, the Spirit calms.
- Spiritual Properties: balance and alignment of physical and metaphysical forces, alignment parallel, promote mysticism

When used during healing sessions, the therapist is a stronger listener, showing kindness instead of sympathy. This also promotes active listening, concentrating on what is said during a healing session, and open the healer's spiritual domain to radiate unconditional love and acceptance through the Heart Chakra. You will start to understand that, working with this beautiful and delectably crafted crystal, you are once again free to open your heart to love and trust your sense of peace with your Divine Source.

ANGELITE

The foundation of consciousness is the cornerstone of the Aquarian Age, the New age for spiritual growth, self-acceptance, and others. It is a very recent finding, particularly in Peru; it is also present in Egypt, Mexico, the United Kingdom, Germany, Poland, and Libya.

The colors, from transparent and colorless and brown, gray, and variable shades of blue and violet, cover a wide spectrum. The color of Angelite is a light bluish-gray. This is most commonly found in its natural state as a tumbled stone.

Angelite calms broken hearts, instilling a quiet sense of calm and tranquility. This will encourage the healer to deepen self-affection with the Divine Source, increase intuitive perception, inner awareness, and increase body experiences. Once worn on the human body, a sense of calmness is strengthened, fear of the unknown is dispelled and the senses are calmed.

Gifted to us all by the harmonic convergence in the late 1980s, this element facilitates telepathic contact among two individuals, who are profoundly spiritually linked and among individuals in intimate physical relationships. This will strengthen the spiritual bond between people who are physically separated.

Angelite is associated and resonates with the Chakra Throat, the Chakra Crown, and the Third Eye. The gentle, free-flowing energy intensity will dissolve all blockages in your Throat Charka due to unresolvedrage, fear of your truth being spoken.

This encourages reconciliation of the Self and others, freeing those involved from what is behind you now. Place this item on your third hand, and brace yourself before being "transported" to the Angelic Kingdom, the Kingdom of Etherine Unity.

Put Angelite on your head, symbolizing your crown chakra region, to expand your knowledge of your divine link with the Creator of All That Is.

This precious stone can be put in a glass of filtered water, leftover overnight to fill, re-aligning the molecules of water and drinking that water will create an inner equilibrium, inner harmony, and feeling of internal joy.

DIVINITY ASSOCIATION WITH THE ANGELIC REALM

Archangel Haniel, meaning word, Inner Wisdom, and Power, will allow you to express yourself with love, humility, and sincerity. Archangel Gabriel, whose name meansthe Messenger of God, is going to help you to speak your truth and accept who you are.

- Healing Properties: restores physical body, calms feelings, dispels rage and deceit, unleashes fear of the unknown
- Vibration Frequency: Free-flowing, gentle, subtle, free-flowing, quiet, peaceful place.
- Spiritual Properties: insight, the extension of divine knowledge, astral journeys, heightened internal consciousness, and direction.

Angelite is a blessing to be valued, a blessing from the World of Nature to those who are open to the influence of the leading physical realm.

RHODOCROSITE

Devotion and emotional healing are frequently associated with rose. Emotional recovery is an important part of everyone's path. It exists only love or fear. Love is greater than fear-based thinking about deprivation, doubt, and despair. Love heals. Love is a blessing;fear consumes our strength.

- Think of emotions dependent on devotion: happiness, unconditional love, self-love, love of others, complete participation, freedom of judgment, peace, harmony, prosperity, serenity.
- Thought or emotions focused on fear: uncertainty, doubt, rage, disappointment, rejection, judgment, conditional love, loss, hate, envy, jealousy.

Everything is greater than love. There's nothing more than true love, complete acceptance of you and others with all the flaws, nicks, dings, and dents. The main thing is to give up the fear of being unloved, unlovable. Now you are complete and fine. You are cherished to the fullest.

The pink and white layered stone symbolizes the many dimensions of our inner feelings, the core of our Emotional Energy Body. White and pink striations merge and align heart and mind, soothing past and current wounds that harm the spirit.

Rhodocrosite is precious for its outward and internal beauty so that every one of us can carry out our inner desires for happiness, to give, and to receive, to love unconditionally.

In this stone, the beautiful layers of white are capable of helping you to direct or to access sacred white light filled with sacred healing love.

The solar plexus and the heart chakra both invoke a specific yet strong frequency, which accommodates each chakra that spans the lower physical realm and desires the higher spiritual Self. Rhodocrosite and malachite work together hand in hand, unlock solar plexus energy blockages, remove emotional obstruction in the Heart Chakra.

Once worn or positioned next to the heart chakra, the healing cycle from heartburn, isolation, depression starts slowly and gently, it promotes self-love, it opens the heart to love again. Forgiveness is the secret to redemption in all areas of space and time on all levels. Be ready to release you, freeing you from everything you foresee on your journey.

Are you among the talented ones who receive messages in your dreams from the angels or your spirit guides?

Slip a few Rhodocrosite bits under your pillow or at your bedside table. Put this pillar in a glass of filtered water so that for at least 12 hours the water absorbs loving energy before drinking.

DIVINITY ASSOCIATION WITH THE ANGELIC REALM

Archangel Jophiel, the name means, Beauty of God, the Archangel of Unconditional Love, who sees through the eyes of faith, gradually and profoundly gives us her heart.

- Healing properties: heartburn, loving thoughts, loving relationships.
- Vibration Frequency: high energy, but gentle, strong, caring, safe vibration.
- Spiritual properties: human, physical, mental, and spiritual bodies combine, intuition increase.

Being prepared to fix confidence issues and feelings of fear helps you to release the past wounds and open up the path for a future full of love and compassion. Yes, you will encounter unpleasant experiences while traveling the road. Know that you have the tools to remove and dissolve obstacles.

Autonomy, self-awareness, and self-expansion are among the many gifts given Kunzite's unique vibration. It is smooth, strong, and auto-regulating in strength that best suits the consumer. This is a stone of Unconditional Love that heals deep emotional wounds of the past.

It is usually pastel-pink or soft lilac-pink, the crystalline shapes of which seem to have many small, vertical ridge-like lines that act as piping or energy channels. Kunzite is a mighty healing stone that calms the heart, softens the heart's emotional pain and rejection.

This will make you expand your connection to the Infinite Source of Light and Love, the Source of Energy, which clears up your link with your inner self, your essence, and the seat of your emotional existence.

This gemstone integrates the Heart, Throat, and Third Eye Chakra energy centers if you are willing to undergo such an emotional transformation. Be prepared to block your Throat Chakra, allowing you to speak more lovingly to yourself and others, encourage Heart Chakra to heal at its core, and enable you to see through your Third Eye beyond sorrow and pain.

First of all, be compassionate when you work with Kunzite. In time, it will allow you to concentrate on positive thoughts, and connect with gentle words and with the tone of your voice. Love is an unbelievably purifying energy, which penetrates your cellular physical and emotional body. You can start to experience peace in a very short time.

Once used during mediation, Kunzite centers the physical body, helping to coordinate Heart and Throat Chakras in perfect harmony with each other. This amplifies a sense of moral harmony.

DIVINITY ASSOCIATION WITH THE ANGELIC REALM

Archangel Jophiel, whose name means Beauty of God, if asked, will allow you to radiate compassion, to offer and receive love unconditionally, without expectation.

- Healing properties: intense mental recovery, nourishment, sympathy, relieves depression
- Vibration Frequency: quick, heavy, powerful self-regulating energy that integrates internal and external self
- Spiritual Properties: strong relation to Divine Energy Source, allow higher consciousness

Allow Kunzite to protect your Aura from unnecessary adverse energy, mental assault, and promote constructive self-discussion, self-love, and healing.

Be able to treat at an accelerated pace. Be able to link to your higher consciousness. Be prepared to release everything that you have thought about or believed about love, whether family, friend, or intimate love.

Be prepared to remove the unpleasant and traumatic experiences that no

longer require your attention to unlock the road before you.

LAPIS LAZULI

It's a strong stone known as the stone of communication that lets you express your truth with honesty. It increases a sense of peace, calms broken nerves, and increases understanding of oneself. The blue color ray symbolizes movement and free-flowing energy.

Lapis Lazuli's color is dark blue with gold flecks. Ancient Egyptians appreciated this lovely gem because of its inherent ability to link both it with the physical and the gods' realms. The soul was thought to be in the mind, in the middle of the brain.

This was used during meditation so that the user would relate more completely to the True Root of All There Is, the Kingdom of Gods. This was believed to be a spiritual healer and purifier of the soul. The dark blue and gold were symbols of dominance for the Pharaohs and the High Priests and Priestess.

The Throat Chakra and the Third Eye are resonating with Lapis. When worn or used as a pendant on the Throat Chakra, it will begin to dissolve and eliminate all barriers, allowing you to express your truth with clarity and compassion.

Placing Lapis on your Third Eye Chakra in the meditation cycle will purify and stabilize your energy center and clear the way for views outside your understanding of the physical world.

This golden blue gem symbolizes positive thought, heart innocence, infinite possibilities, and the wisdom of the ages. Holding, wearing, or putting Lapis next to you will help you unlock restricted convictions and push you into the road ahead. It promotes integration with your Etheric Energy Body between your mental, physical, emotional, mental, and spiritual bodies.

DIVINITY ASSOCIATION WITH THE ANGELIC REALM

Archangel Taharial whose name means the Purity of God, will help you talk with honesty, dignity, and the careful decision of your actions so that no harm can be done to yourself or others.

Archangel Gabriel, the name is Messenger of God, will help you remove blockages in the Throat Chakra, gracefully and comfortably embrace yourself.

- Healing Properties: self-expression, dignity, harmony, and purification
- Vibration frequency: quiet, powerful, intense blue true alchemy and communication
- Spiritual Properties: age wisdom, pure mind, heart pureness, soul cleanser.

Working with Lapis Lazuli for a longer time, the directed physical world energy can be converted into the higher vibratory energy of the spiritual realm using the golden spots in its core essences.

A word of warning, if you are not yet ready or able to embrace your reality, your inner self, then it is better to wait for you to be sure you are ready before starting.

BLUE LACE AGATE

Blue's energy level is soothing, flowing, relaxing. The color blue is related to the water dimension, which means feelings, strength, tranquility, serenity, and harmony. Although it's subtle and gentle, it can be quite tumultuous and stormy at times.

In times of emotional turmoil, remember to embrace and respect the "current condition" or room you find yourself in. To calm broken nerves, emotional wounds, and light a blue candle, concentrate for some moments on the flame, which will encourage your whole being to relax. Feel your shoulders release everything they're holding.

Blue is the throat chakra color, contact of loving-kindness, expressing your truth. You should speak caring words. There are many ways to express what needs to be saidwithout harshness, without resentment. Indeed, sometimes it is not possible. In situations where you don't speak to yourself or to others kindly, forgive yourself and move on. Every one of us does the best we can at all times. We are all a "work in progress", including me.

Join me while we travel through January, as we float along this path from Blue Lace Agate, Kyanite, and Lapis Lazuli to Blue Calcite. Take this journey of renewal and regeneration.

Blue Lace Agate,popularly known as the Wishing Stone, emits warm, peaceful energy. Used during meditation or worn on the body, it increases the feeling of joy, calms the emotions, calms the nerves.

This stone is linked to the water dimension as its beautiful blue and white striations symbolize. Just like water flows from one location to another, let your emotions flow easily. Water is most often colorless, but it changes its shape to match the current boat.

The stone will resonate with the Throat Chakra, and it will allow you to speak your truth, without hurting you or anyone else. Blue Lace Agate helps your Throat Chakra to remove the barriers created by your thoughts, feelings, and disappointment with yourself and others.

This will clean up the Auric energy field and stabilize this. Place a glass of filtered water with some tiny tumbling stones; leave it overnight; drink the energycharged water in the morning. The charged gemstone water blends with your inner self and radiates the free flow of light and love all over your physical body.

The soft blue energy helps the consumer on the journey ahead, dissolving unnecessary negative energy habits. Working with this stone will help you avoid harmful patterns of actions. The subtle frequency of energy can be used to improve intuitive inner perception.

Allow yourself to interact with circumstances as they unfold around you. You are equipped to show compassion and inspire not only you but those around you.

DIVINITY ASSOCIATION WITH THE ANGELIC REALM

Archangel Gabriel, the name means, Messenger of God will help you express your truth with peace and compassion.

- Healing Properties: calms the passions, calms the nerves, dissolves the Throat Chakra blockages
- Vibration frequency: soft, subtle
- Spiritual Properties: increasing perception, inner vision, strong vision.

Blue Lace Agate will help you travel with grace and ease along the lane, radiating light, and love with a sense of peaceful calm.

RED CORAL

When you look at the underground garden of Coral, between the many fishes, you will glimpse a siren thathides her rainbow-colored underwater treasures!

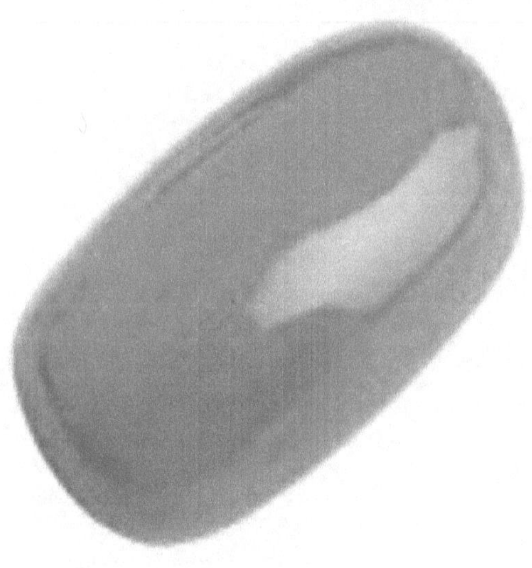

This stone is found in the vast gardens under the majestic waves where you can find spectacular coral reefs in countless varied colors. Red Coral is bright, powerful and subtle, all at the same time as if its special energy could be controlled by magic.

Red Coral resonates with the Kundalini Energy, Root Chakra, and inner vision and helps to dissolve inner blocks of energy. It facilitates a subtle but profound vibration purification which enables the consumer to improve his inner vision. Working with this incredible marvel in the sea dissipates excess nervous energy and moves you closer to the loving energy of the Earth Mother.

Coral was highly considered as a guardian in the underworld among the many practices of the Ancient Egyptians. High Priests and Priestesses also decorated necklaces when they returned from the land of the Nether. It was claimed that every piece of Red Coral held a drop of blood, the Sacred Blood of God and Goddess inside its many small chambers.

This wondrous gem of the sea can be used in meditation to activate thoughts and emotions that allow the mind to peacefully obtain inner guidance. The original people, the indigenous people of this land, also consider it one of the four basic stones.

The deep organic bond to Earth Mother helps the wearer to increase his inherent intuitive and visualizing abilities.

Red Coral symbolizes peace and joy, success and richness, and security from a psychic assault, which is a negative thought or other force.

DIVINITY ASSOCIATION WITH THE ANGELIC REALM

Archangel Raguelwhose name means "Friend of God" will allow you to loosen self-made limits, expand your boundaries or comfort zone.

- Healing properties: emotional calm, mother-child bonding, relaxed feeling.
- Vibration Frequency: intense, fire, grounded, speed of burning.
- Spiritual properties: increased meditation, activation of a third eye, clarifier of intuition, protection against psychic attack.

The bond between Mother and Child can be strengthened if everyone wants to improve his relationship by carrying or wearing a piece of Red Coral.

CINNABAR

This very enigmatic stone's name includes many names or markers. It is known as the Stone of the merchant, the Stone of the Alchemists, and Dragon Blood of ancient Persians because of its rich vermilion color.

In the Orient, Cinnabar is desired for its capacity to give long-life to the wearer. Delicately sculptured beads and bracelets are embellished with Mandarin Duck sets. Mandarin Duck is among a few who live, symbolizing pureness, long life, trustfulness, and lasting love between partners.

This is called the Stone of Alchemists because it contains traces of Mercury, which was once believed to be gold. The stone of the merchant is so renowned for its ability to draw success with honesty.

This will also empower you to develop interpersonal skills, minimize uncertainty, and build a sense of teamwork and collaboration while working with colleagues or collaborators.

A Foundation or Root Chakra, the Sacral Chakra, and the Third Eye overlap in Cinnabar. Your Basic Chakra is connected to physical and financial survival. Responding to the sacred or sexual chakra, you will tap into your imagination and discover secret gifts of inner speech. It will help to restore love relationships and instill a sense of trust.

The healing properties help to transform thoughts, emotions, and to remove or dissolve self-worth and self-esteem barriers. Carry a cinnabar slice in your pocket or coin pocket. Put some on your desk or money corner.

Take time to work with this powerful stone, do not rush or force yourself to fuse with the high intensity of its energy frequency. It will promote the connection of the physical realm with that of the spiritual realm.

Start with meditation, gradually increasing the length of it. Place it over or on your Third Eye for the elimination of retained or unused energy, enabling you to distinguish the many choices ahead of you.

DIVINITY ASSOCIATION WITH THE ANGELIC REALM

Anauel, an angel of wealth and commerce, will help you handle your business ties and help you solve money and finance problems.

- Healing Properties: energy blocks, Chakras alignment, doubt, self-worth, presentations, company and finance problems.
- Vibration frequency: solid, cured, purifying.
- Spiritual properties: defend against other people's negative feelings; mediate assertive or hostile behavior, spiritual growth, transformation.

A major advantage of working with Cinnabar is its high vibration frequency

at which abundance and wealth are refined, purified, preserved, and attracted.

Carved jewelry is often seen as a symbol of commitment, love, and long life.

MALACHITE

Malachite is one of the oldest known stone, a hard, healing stone that calms the nerves. This is a strong stone to bring prosperity and success.

The ancient Egyptians used the strength and regeneration of malachite in their bodies. This many-layered stones helps to ground nerves or excess energy, enabling the Crown Chakra to more completely open and allow higher energy levels to be channeled.

Malachite color is present in many shades of green and produces a smooth, fluid effect when viewed. This is a heavy stone that is both normal in form, polished on one side, or tumbling stones.

Instead of releasing carbon, Malachite consumes energy that transmutes energy in itself. Place this stone over areas that might be uncomfortable, it will remove from you lower frequency energy, which causes discord within the physical body.

DIVINITY ASSOCIATION WITH THE ANGELIC REALM

Archangel Raziel whose name means "Secret of God" will help those who seek to align themselves with their Higher Selves and dissolve blocks within their physical and mental bodies.

- Healing properties: overall health; absorbs energy, helps resolve barriers in physical and mental bodies.
- Vibration frequency: thick, strong, very polite to use.
- Spiritual properties: channel higher frequency vibrations that enable you to easily connect to the Spirit's guides.

Use this stone as an instrument to focus your energies, to help Mother Earth restore herself, and to heal deeply. The striations, the curves, the Malachite layers are part of his beauty and the very essence of this beautiful stone.

Malachite is best cleaned and recharged by placing it in the sun or on a large, clear quartz cluster.

Emerald

The Emerald is seen as a link between mental and emotional worlds, a link between two individuals, and a desire to give and to receive affection. This has no ability alone to create a healing effect on the human body. But it provides the user, recipient, and healer with a deeper energy pulse, with a concentrated purpose, to facilitate healing.

Additional qualities include improving perception, self-forgiving feelings, and others. It strengthens and extend what is learned, and bridge the influence of concentrated thought and mental strength.

The Emerald is closely associated and resonates with the deep Heart Centre, which influences emotional healing. Ancient Egyptians claimed it would protect the traveler in the left arm; it reflects unconditional love, fellowship, relationships, a sense of unity.

The color varies from deep green to light green, almost green-yellow. The power signature and the vibration frequency are capable of being gentle, soothing, caring, giving, strong, subtle, and free-flowing.

DIVINITY ASSOCIATION WITH THE ANGELIC REALM

Archangel Jophiel, whose name means "Beauty of God" is going to make you see through love's eyes; beyond surface appearances beyond what has been portrayed.

- Healing properties: compassion, affection, determination, ability to alleviate back pain.
- The vibration frequencys: delicate, powerful, subtle, free-flowing
- Spiritual properties: the capacity to develop intellectual wisdom; extends the gift and reception of the heart and mind.

It's best not to always wear an emerald, because it could overestimate mental activity and potentially cause sleep deprivation with an abundance of thoughts.

Emerald may appear faulty, but do not considerit a lower gem because its curative properties are not yet fully understood. Be careful with yourself as you work and wear this beautiful green gemstone.

Citrine

The Solar Plexus Chakra or energy core just above your navel resonates with citrine quartz. Citrine is light golden orange to dark orange and in some specimens almost gray.

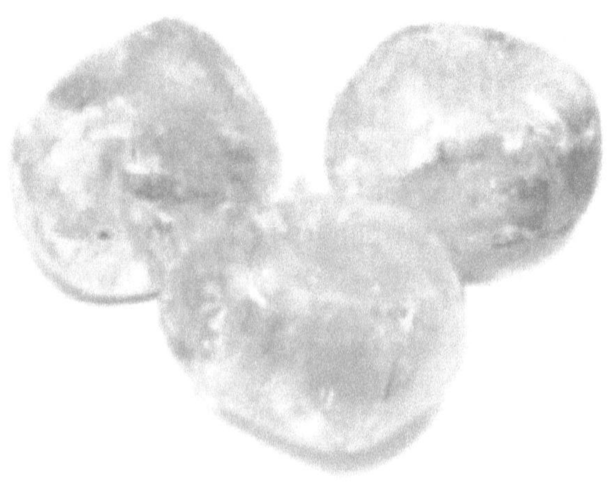

Citrine can be present in clusters of very small sizes to larger sizes. Citrin can also be present in both ends of the crystal structure (double termination points) and single points.

The frequency vibration of Citrine Quartz can help you to focus on what you want in your life. Reflect on what you want rather than what you don't want when you center your thoughts. You can keep your citrine cluster or place it in front of you during your meditation practice and concentrate your eyes on the cluster.

If you want to use Citrine to liberate you from the past, clear your mind and concentrate on what you want to do. Let the Citrine's energy pulse enter the brain until it is gone into its soothing warmth.

Citrine quartz clusters can be utilized in the strength of the body when positioned on the Solar Plexus (above the navel), the Heart Chakra (in the middle of the chest), and Crown Chakra (above the head).

When stored in sunlight for many hours once a week, citrine is highly energized. When you use other crystals to improve total energy usage, Citrine may not have to be washed or recharged every week. Use your intuitionto guide you.

DIVINITY ASSOCIATION WITH THE ANGELIC REALM

archangelUriel, the Angel of imagination, creation, and prophecy whose name means "light of God" helps you to state or write down the highest and greatest benefit for those involved with pureness and honesty.

- Healing Properties: relieves depression, increases self-empowerment, facilitates self-expression, imagination, stabilize the nervous system.
- Vibration Frequency: dry, soothing, energizing, penetrating.
- Spiritual properties: frees the self from the past

AMETHYST

Amethyst is referred to as the Jewel of the Healer and the Jewel of the Heart. Amethyst is resonating with the Crown Chakra right above your ear. The purple color is often associated with the third eye chakra in the middle of the face, right above your eyes. The Amethyst color spectrum is very light lavender to almost dark purple-blue. Amethyst energy calms the mind and spirit, improving the experience of meditation. Amethyst can be present in clusters, dots, and walls.Big Amethyst clusters may be used for purifying and recharging smaller crystal points. Try at least once a week to sweep up, arrange, and refresh your Amethyst clusters. Amethyst clusters can be exposed to direct sunlight for a long time.

Clean and recharge the Amethyst overnight by the Moonlight or by the candle flame (use caution as extremely high heat can be created by the fire of a candle).

The frequency pulse of Amethyst is calming and relaxing, allowing you to extend your consciousness, raising the veil between the spiritual world and the physical world. During meditation, amethyst may be found directly on the

Third Eye and the Crown.

It is recommended to keep an amethyst point or a small cluster on your left side so that your body will respond to the vibration of Amethyst. When your work with Amethyst is ready to be expanded, put Amethyst right on your Third Eye or Crown Chakra.

You will know when it is time to continue working with Amethyst on the Third Eye and Crown Chakra. Do not hurry and be patient when working with Amethyst, the soft vibration is very powerful and, at first, it can be overwhelming.

Amethyst can be worn or kept in your bedside table or under your pillow to help relax your busy, overactive mind to bring a more relaxing sleep. While holding Amethyst, the end should always face or point to your chest, allowing the energy of Amethyst to flow in and through you. Amethyst and Rose Quartz can be used together in heart and mind self-healing.

DIVINITY ASSOCIATION WITH THE ANGELIC REALM

Archangel Raphael whose name means "God heals" the angel of healing and peace, assisting the Holy savior in every one of us, protector and guide of healers, patron of travelers.

- Healing properties: Healing, decreases mental stress, can help relieve migraine headaches.
- Vibration frequency: soothe, relax, purify, cure, awaken, turn
- Spiritual properties: modesty, spiritual awakening, enhancing the connection with the spiritual realm.

CHAPTER 5. ACTIVATION FOR MANIFESTATION

Crystal grids are tools to help you concentrate your energy and set your goals. You may put a picture or photograph of your heart desire in the center of the grid.

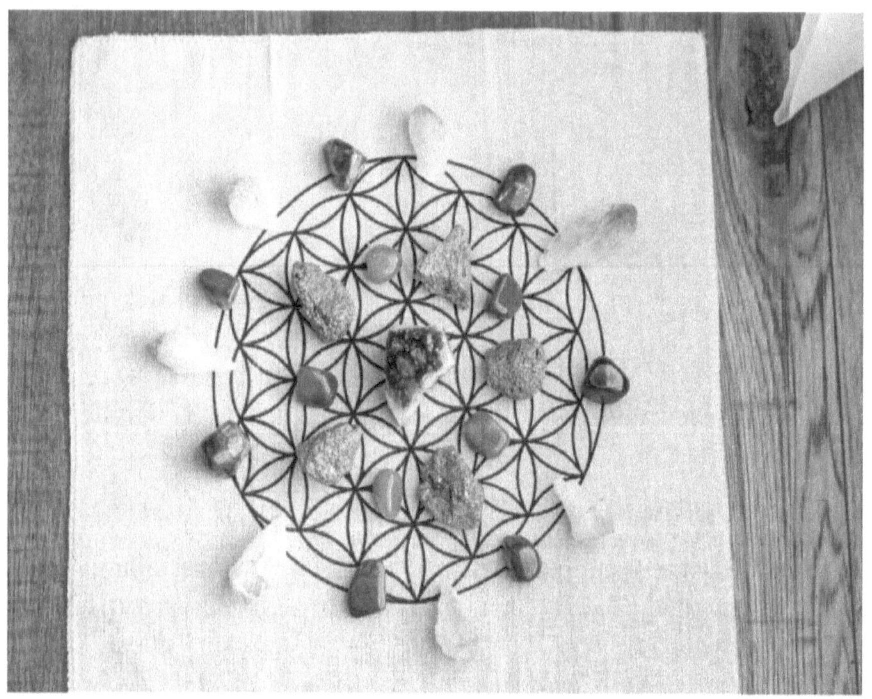

The key feature of a crystal grid, when triggered, is the creation of a beautiful visual reminder of everything that you want, the wishes that you set up, and release them into the world for physical manifestation.

With this chapter, we will explore how an Agate Crystal Grid can be generated and triggered. You may choose to use this form of crystal grid or build one of your own. The grid has six small crystals or stones at its edge and a bigger crystal or stone at its middle.

Select crystals or stones to which you are attracted to or feel perfect with. After you've made your selection, clean and charge the crystals before placing them on the grid. You may choose to build your sacred space before activating it if you did not. To trigger your crystal grid, you will need about 20 to 30 minutes.

Place the six smaller crystals on the agate slice around the bottom, placing the bigger crystal in the middle. Fix the crystals on to the agate piece.

Be still, breathe deeply, to activate your grid.

Slowly exhale... Inhale... Exhale... Inhale... Exhale... You can choose to play soft music, light incense, or candles. Start by looking at or reflecting on the central crystal, visualize, or imagine the energy of white light from it.

When you concentrate on the central crystal, start to trace the path from the center of the crystal to the closest crystal, just above it, and turn right to the next and back to the center.

Take a few minutes to imagine the white light in the shape of a triangle, hold on until you see or hear the central force. Based still on the center crystal, a track starts from the center to the second crystal, to the next or third crystal, and back to the middle of the crystal.

Keep your attention until the white light energy from the first two parts of the Agate Crystal Grid is sensed or heard. Then, proceed in the Agate Crystal Grid until all six grid sections are activated.

Selenite is one of just two minerals that regenerates energy and requires no cleaning. This crystal cures the user at the cellular level, removes past-life problems so that you can continue on your path. It will allow you to remove yourself from the past.

Often separation is required to move forward. It's not all about separating from your friends, family, or partner. It's about freeing you from thoughts, feelings, experiences, which generate lower vibrations of energy and make you lose sight of your direction.

Selenite Blades are excellent tools to cut negative energy wires, remove mental uncertainty, and explain every situation. A blade can also help you expel blocked or stagnant energy from the inside. A vacuum is created by eliminating static energy, allowing positive White Light into your physical body, your aura, and your etheric energy fields.

The Crown Chakra is most closely linked to this incredible mineral specimen, as it enables discernment, deepens the perceptions of contemplation by delusion, facilitates the opening of the higher Chakra energy centers, connecting you more directly with the Spiritual Realm.

Healers may benefit from hitting the distance through time and space. A Selenite Wall or Blade may be used to direct healing energy to a specific location within the physical body and past life.

Selenite can be used for constructing a security grid around your home by placing a blade or wall on each of the external corners. Place a few inches below the surface in the field with the point facing up.

You have the choice to completely bury the Selenite or allow it to be only partially buried. Both methods are perfect; choose the method that resonates with you.

When you become more accessible to the Spirit world, you bond with your spiritual leaders and angels in a deeper way.

CHAPTER 6. THE ART OF CRYSTALS CLEANSING & EXPLORATION

Why must we purify our crystals? There are two things; first, we need to clean our fresh and exciting crystals from actual dirt and dust.

Fresh crystals also have dust on them, which must be cleaned from the store or market stall. Secondly, we must also clean it vigorously. What does that mean?

You must remove the energy that others left on the crystal because it was already handled by many other people when you bought it.

There are 8 ways to clean your crystals which are mentioned below:

1. Air: One of the easiest methods of immediate cleaning, when

you buy your crystal, is to hold your crystal in your hand, and then blow your breath on the crystal gently to clean it. When you do, you apply your energy or prana to the crystal and this eliminates or negates any other person's residual energy; of course, it does not eliminate actual dirt and dust.

2. Water: Under flowing water, most crystals can be cleaned. Caution: crystals can dissolve in water. Some examples include selenite, cyanite, azurite, or any of the soft crystals. You should clean the actual dirt with a damp cloth. Some books advise you that you have to use seawater or spring water, but that is often not practical, and you can use the purest water with little to no chemicals. It is also best to wash physical debris off all your crystals daily. It is often overlooked if you use the cleaning method of sunlight or moonlight.

3. Moon Power: Put your crystal under a full moon outside to infuse it with the moon's energy. Moon energy is soft feminine energy and it is a very gentle form of purification. It is also a safe idea to cover the crystals with a net so that they are not taken by the night animals.

4. Earth:Another common way is to bury your crystal. Some people claim that this is the best way of purification. This method includes a few measures, note where you hide them, and do not hide them too deeply. Often, you're never going to find them again because Mother Earth recovers them. One good way is to bury them in a pot so that you can easily find them. You will need to clean them after you dig them up because they will be dusty.

5. Smoke: White sage has been used for many years for cleaning native crops and it is a highly efficient way to clean multiple crystals simultaneously. If your collection is huge, it would take you all night to get it to the moonlight. You can smudge them with a white wise stick or white incense. Smoke over your crystals and lighten your incense or smudge stick. The incense of sandalwood is also a good cleanser. If you don't like

these fragrances, almost any incense helps purify your crystals. You can close the door to a room with resin and charcoal blocks and purify the whole room at once.

6. REIKI: You can use Reiki energy for cleaning crystals if you've been tuned to Reiki. When you are a Reiki level 1, keep the crystals in your hands and give Reiki to your crystals deliberately.

7. Additional Crystals: Flat amethyst clusters are perfect to use with crystals; they are energy-cleaned just by placing the crystals on the amethyst for a few hours. Geodes and clusters of Transparent Quartz are also very useful.It doesn't matter which tool you use to clean your crystals. The most important thing is to clean all your crystals regularly both physically and energetically.

Naturally And Intuitively Exploring Crystals

The goal here is to explore the crystals and to discover their energy and potential. You can have something in particular, such as a physical, emotional, mental, or spiritual problem or ailment that requires help.

Your goal may be to increase your intuition and intuitive capacity, communicate with your superior self and guides, and ultimately discover your spiritual self. No matter what your target is, it is infinitely easy to approach your crystals!In order to explore your crystals intuitively, then you must:

1. Agree to approach your crystals and work with one, no matter what your purpose is (at this stage you have established your purpose).
2. Pick any crystal that is at that moment appealing to you without question or searching for details (the appropriate crystal / s would appeal to you so depend on the process).
3. For a moment, you hold that crystal and decide how to use it.

Perhaps you would set this goal already at phase one if, for example, you want a crystal to take with you into the bath or want a crystal to meditate with.

4. Whatever system and crystal you use, you know that this crystal helps you. Follow your gut as to when you should put the crystal back, etc., you will never go wrong and that is easy – reach out and listen.

When you choose to use your crystals in some way, your energy reaches with or without a particular intent, and you choose a crystal that resonates with you. Remember that crystals can benefit us in many ways-by increasing or relaxing energy, protecting or grounding, restoring balance/healing, etc.

Some generate a feeling of encouragement or inspiration; some help you to feel safe, less afraid, and more confident; some draw attention to emotions or pain that you need to cleanse; some strengthen positive feelings to help you cope and push your life in the right direction.

Anyway, unbalanced crystals and disruption can help to restore your health and concentrate. Whatever aspect of your life you want to focus on, crystals will help to improve, kick-start, and sustain this cycle. Regardless of the purpose, simply choose one or more crystals and let yourself be drawn to them.

Many people can see or smell the crystal in some way when they use it. This would typically be better at times when you are profoundly relaxed and hence more receptive, for instance in meditation, sleep, and dreams, or when you relax with stones in or around you, etc.

To others, the results will only be felt later as they understand the changes taking place in themselves.

It may be a shift in outlook or mood, a feeling that you are happier than in the present rather than in the past, that you have more energy and positive attitude, and most importantly, that you feel content inside yourself and connected to the larger picture so that you travel more intuitively through life and have more relation to your true self. Is this not the goal anyway?

As we link internally to discover who we are, everything else in life comes into being. If at this point you feel light years away from this self-discovery, ask your crystals to help you, but you also have to take some time to achieve quietness in life or you will stay busy and distracted and avoid yourself.

Explore with Silence

Start by spending time sitting still with a crystal and exploring it. Take a look at it, concentrate on your breath, let your mind float away, and let yourself become one with this stone. The use of crystal as a focal point makes it easier for the mind to remain calm than if you only close your eyes and construct all sorts of thoughts and pictures.

So, focus on the stone, see the edges blurring and let it go blurry. Only let it be. Don't be annoyed with yourself if your thoughts wander away, simply return your attention to the crystal and your respiration and start again.

Doing so for 5 minutes a day will help you begin an incredible journey both personally and spiritually. Because we are "active minded" people, it's one of the most difficult things to relax the mind, but once we have a whole new world of possibilities, something that gives us a lot of joy.

Never again can you look at yourself or the world the same. But I want you to

explore that, and I wish you all the best!

CHAPTER 7. FREQUENTLY ASKED QUESTIONS

Crystals are beautiful, powerful, and fascinating. They have supernatural abilities and talents. The crystals can be produced, processed, modified, transmitted, and transformed for the monitoring and activity of the radio and crystal laser. The "positive vibrations" crystals harmonize the atmosphere or the body.

Crystals can be programmed to send "good vibes" to your atmosphere and are suitable for the improvement of your house, car, or workplace. You attract money, joy, friendship, and everything in your life.

The ability of Crystals to absorb energy ensures that they can be used for other activities such as converting healing energy into an emotional impediment or a point on the body. The future of crystals and more is divine.

By using the exquisite strength of the crystals, you can change your life. When you know how to gain strength, you will get you all your heart desires. Crystals are also used for curing.

In this sense, healing doesn't mean 'to improve,' it means to improve your health and make you feel good, although in ancient and medieval times crystals had been grounded and administered.

Have our forefathers influenced the use of crystals today?

Absolutely. For thousands of years, crystals were not only used to decorate, but also to cure and influence life. Ancient people believed that crystals were the gifts of gods and contained the spirit of a goddess. Amber beads were, for example, found in tombs more than 8000 years old, and amber is still today a protective stone.

In ancient Mesopotamia around 4000 years ago, some birthstones were used at that time to attract gods and to cure iron-containing hematite, sacred in Mars, and used to treat diseases in the body, just as now.

Crystals were an integral part of ancient Egypt's medicinal practice. The

Vedic gems associated with planets in India were used for at least 3000 yearsand probably even longer.

When do people discover the magic of crystals and gemstones in their everyday lives?

Crystals come in all forms and sizes. They are all brilliant, beautiful, and often costly. Some are hard bits, obviously uncomfortable before you know their secrets. A diamond or ruby can be ignored in its raw state. Most stones are tumbled, cut, or scratched, but in their normal state they function the same and you can easily put a crystal into your pocket or pillow and allow it to do its job.

The magic of a crystal must be magnetized to your strength to work. This aligns the crystal with your intent. It helps focus on what the crystal has to do and guarantees that the crystal does its job.

The engagement of a crystal significantly improves the crystal's output and ensures that is used fora good purpose. Take it around, take it to the tap to purify your energies, then keep it all in hand and schedule it to respond by concentrating on your particular intent.

Passion radiates from many romantic or deeply sexual crystals and stones, including Red Jasper. They attract a soulmate, or they have a relationship with one another. Crystals change the way you look and open yourself up to love. You can feel the love radiating from various crystals, and these stones can be calming and carried through your heart.

They make a gentle effort to make you take up and enjoy yourself – a precondition for the love of someone else. One of the fun ways to find crystal love is by floating with your favorite stone. Pick one that exudes the virtues of your heart: ambition, grace, self-love, for example.

Clean the crystal and apply a few drops of rose oil to it in the cold. Glide the stone under your pillow in the evening to reinforce it.

Any other hints for first timebuyers of crystals and gemstones?

It's the crystal that speaks to you. A crystal doesn't have to be costly or rare to thrive. Volume and size are of little concernin terms of crystal intensity. Don't be deluded by the shiny, sparkling, and immense ones. Those aren't always the most efficient.

Tiny, malformed, and unwanted crystals have the same potential and far less cost. See what crystal first catches your eye when you visit a crystal shop. Perhaps it'd be the one for you. When you are searching for a specific stone, immerse your hand in a pool and your name will be on your side.

Keep in mind that everybody is special — we all radiate different energies, different personalities, and different family rhythms. There is no magical 'curing' crystal (although different varieties of quartz come close!).

What works for you won't work for your partner automatically. Don't be afraid to learn intuitively how to use crystals rather than what somebody else thinks you are doing. After doing that, I've written a series of books to help you find the best way to do it.

Also, be aware that many relatively simple illnesses are not 'dysfunctional.' There is an underlying physical, mental, or spiritual illness that is a disease that attracts your attention. Crystals solve these underlying problems carefully and bring you back to a holistic balance.

One thing is highly important, crystals absorb adverse energy and must be periodically cleaned up. The easiest way to do this is to keep them under running water for a few minutes and then recharge them in the sun.

Are crystals and gemstones recognized for their brilliant benefits?

Of course, I believe that the success of my books, such as the Crystal Bible, shows that more and more people appreciate crystals' value. Today, almost every city in England has a crystal shop and they all say that people are searching for a stone to take their pain away, attract love, etc. so we assembled crystal prescriptions and a crystal healing kit that covers every imaginable phenomenon with healing stones.

What does Conscious Awareness mean?

It's being conscious of what you feel and respect your feelings. When you respect where you are, it is easier to recover, to release emotional energy that no longer fits you best.Lepidolite radiates Earth Mother's gentle calming vibration into a strong mix of gentle and pure, white, mixed with lithium mica and white turmaline.

The emotional and mental energies are intertwined and interwoven. For you, what does that mean? It means that when you feel pain,your thoughts always focus on what you feel.

This seemingly soft stone has a powerful healing vibration that helps you to control your energy bodies mentally and emotionally. It is essential to be able to create and maintain balance in every healing process.

It primarily resonates with the Heart Chakra. This can open and disable all energy centers of Chakra. As you open, trigger, and balance your entire chakras, you resonate with the Sacred Feminine Spirit of the Earth Mother that is the source of all excess emotional energy.

As a transitional stone, Lepidolite helps you to leave everything behind you now, creating a sense of calmness in your heart. You can use Lepidolite to link to Archangel Raphael.

Archangel Raphael will help you heal your heart and restore spiritual equilibrium in your physical, financial, mental, and auric bodies. The Ascended Master is the Divine Healer in you who will allow you to reconnect with your Higher Self.

When you focus on this gentle, nearly magic stone of consciousness, you will gradually become linked with that of the Cosmic Consciousness. Establishing a relationship with the Cosmic Consciousness bridges your past lives so that you can heal unresolved emotional pain and trauma that you might have brought into this life.

Many people are looking for and working on Akashic records. Lepidolite allows you to view your Akashic records. I agree that you can only access your records and that nothing else is allowed without their permission.

Do not let this seemingly simple stone fool you, because its vibration, affection, unconditional love and light, are very strong. It is a marvelous addition to your crystal healing chest and it will help you on your spiritual journey.

Why Should You Consider Using Quartz Crystals For Healing?

Although most of us realize that we can use quartz crystals to heal, our mindset appears to be very cynical or entirely pessimistic. The idea of using quartz crystals to create healing sounds very new, and we know of almost every disease. To most of us, though, the whole thing sounds like just one of the crazy theories promoted by the "new age" movement.

However, the fact is that many people have greatly improved their wellbeing by using quartz crystals. Until we continue to discuss the reasons whyyou, too, should consider using quartz crystals for healing, we will take a quick detour to explain how the quartz crystals are connected to healing.

The true cause of illness – at least according to oriental medicine practitioners – is the misalignment of the body resources. We all know that our bodies are made of energy from our study of primary physics. All is energy, and there is no explanation of why our bodies are an exception.

The news for us is that when our body's energy vibrates/flows correctly, we continue to achieve optimal health. When the energy flow/vibration is disrupted, our wellbeing begins to beat up.

There are now certain substances in our bodies that can restore the right rhythm and flow of energy. In this group there are quartz crystals. So, the mechanism by which the quartz crystals operate is to correct the flow of energy and vibration problems that continue to evolve by various diseases.

You can still have doubts about the ability to heal, but you will have to believe it when you hear testimony from people who have benefited from it, and when quartz crystals are used to cure your own body. There are many reasons why quartz crystals should be used for therapeutic purposes.

First of all, as you believe, the quartz crystals are highly successful in healing. Most surprisingly, the effect of their use is almost instantaneous. Moreover, because the entire process is handled externally, you don't have to get into internal contact with it, they are also better to use for healing.

Finally, you must realize that quartz crystals are extremely cheap and almost everyone can use them in healing-with the right training.